PHYSIOLOGICAL LIMITATIONS AND THE GENETIC IMPROVEMENT OF SYMBIOTIC NITROGEN FIXATION

ADVANCES IN AGRICULTURAL BIOTECHNOLOGY

Akazawa T., et al., eds: The New Frontiers in Plant Biochemistry. 1983.
ISBN 90-247-2829-0

Gottschalk W. and Müller H.P., eds: Seed Proteins: Biochemistry, Genetics, Nutritive Value. 1983. ISBN 90-247-2789-8

Marcelle R., Clijsters H. and Van Poucke M., eds: Effects of Stress on Photosynthesis. 1983. ISBN 90-247-2799-5

Veeger C. and Newton W.E., eds: Advances in Nitrogen Fixation Research. 1984. ISBN 90-247-2906-8

Chinoy N.J., ed: The Role of Ascorbic Acid in Growth, Differentiation and Metabolism of Plants. 1984. ISBN 90-247-2908-4

Witcombe J.R. and Erskine W., eds: Genetic Resources and Their Exploitation – Chickpeas, Faba beans and Lentils. 1984. ISBN 90-247-2939-4

Sybesma C., ed: Advances in Photosynthesis Research. Vols. I-IV. 1984.
ISBN 90-247-2946-7

Sironval C., and Brouers M., eds: Protochlorophyllide Reduction and Greening. 1984. ISBN 90-247-2954-8

Fuchs Y., and Chalutz E., eds: Ethylene: Biochemical, Physiological and Applied Aspects. 1984. ISBN 90-247-2984-X

Collins G.B., and Petolino J.G., eds: Applications of Genetic Engineering to Crop Improvement. 1984. ISBN 90-247-3084-8

Chapman G.P., and Tarawali S.A., eds: Systems for Cytogenetic Analysis in *Vicia Faba* L. 1984. ISBN 90-247-3089-9

Hardarson G., and Lie T.A., eds: Breeding Legumes for Enhanced Symbiotic Nitrogen Fixation. 1985. ISBN 90-247-3123-2

Magnien E., and De Nettancourt D., eds: Genetic Engineering of Plants and Microorganisms Important for Agriculture. 1985. ISBN 90-247-3131-3

Schäfer-Menuhr A., ed: *In Vitro* Techniques – Propagation and Long Term Storage. 1985. ISBN 90-247-3186-0

Bright S.W.J., and Jones M.G.K., eds: Cerial Tissue and Cell Culture. 1985.
ISBN 90-247-3190-9

Purohit S.S., ed: Hormonal Regulation of Plant Growth and Development. 1985. ISBN 90-247-3198-4

Fraser R.S.S., ed: Mechanisms of Resistance to Plant Diseases. 1985.
ISBN 90-247-3204-2

Galston A.W., and Smith T.A., eds: Polyamines in Plants. 1985.
ISBN 90-247-3245-X

Marcelle R., Clijsters H., and Van Poucke M., eds: Biological Control of Photosynthesis. 1986. ISBN 90-247-3287-5

Semal J., ed: Somaclonal Variations and Crop Improvement. 1986.
ISBN 90-247-3301-4

Purohit S.S., ed: Hormonal Regulation of Plant Growth and Development, Volume 2. 1987. ISBN 90-247-3435-5

Wolfe M.S., and Limpert E., eds: Integrated Control of Cereal Mildews: Monitoring the Pathogen. 1987. ISBN 90-247-3626-9

O'Gara F., Manian S., and Drevon, J.J., eds: Physiological Limitations and the Genetic Improvement of Symbiotic Nitrogen Fixation. 1988.
ISBN 90-247-3692-7

Physiological Limitations and the Genetic Improvement of Symbiotic Nitrogen Fixation

Proceedings of an International Conference on the Physiological Limitations and the Genetic Improvement of Symbiotic Nitrogen Fixation, Cork, Ireland, September 1–3, 1987

edited by

F. O'GARA and S. MANIAN
University College, Cork, Ireland

and

J.J. DREVON
INRA, Montpellier, France

KLUWER ACADEMIC PUBLISHERS
DORDRECHT / BOSTON / LONDON
for
THE COMMISSION OF THE EUROPEAN COMMUNITIES

Library of Congress Cataloging in Publication Data

International Conference on the Physiological Limitations and the
  Genetic Improvement of Symbiotic Nitrogen Fixation (1987 Cork,
  Cork)
    Physiological limitations and the genetic improvement of symbiotic
  nitrogen fixation · proceedings of an International Conference on
  the Physiological Limitations and the Genetic Improvement of
  Symbiotic Nitrogen Fixation, Cork, Ireland, September 1-3, 1987 /
  edited by F. O'Gara, S. Manian, J.J. Drevon ; sponsored by the
  Commission of the European Communities, Directorate-General for
  Agriculture, Co-ordination of Agricultural Research.
      p.   cm. -- (Advances in agricultural biotechnology ; 23)
    Includes indexes.
    ISBN 9024736927
    1. Nitrogen--Fixation--Congresses. 2. Microbial genetics-
  -Congresses. 3. Rhizobium--Congresses. I. O'Gara, F. (Fergal)
  II. Manian, S. III. Drevon, J. J. IV. Commission of the European
  Communities. Coordination of Agricultural Research. V. Title.
  VI. Series: Advances in agricultural biotechnology , AAB 23.
  QR89.7.I56 1987
  589.9'504133--dc19   88-3794
                                                           CIP

ISBN-13:978-94-010-7126-0   e-ISBN-13:978-94-009-1401-8
DOI: 10.1007/978-94-009-1401-8

Publication arrangements by
Commission of the European Communities
Directorate-General Telecommunications, Information Industries and Innovation, Luxembourg

EUR 11517
© 1988 ECSC, EEC, EAEC, Brussels and Luxembourg
Softcover reprint of the hardcover 1st edition 1988

LEGAL NOTICE
Neither the Commission of the European Communities nor any person acting on behalf of the Commission is responsible for the use which might be made of the following information.

Published by Kluwer Academic Publishers,
P.O. Box 17, 3300 AA Dordrecht, The Netherlands.

Kluwer Academic Publishers incorporates the publishing programmes of
D. Reidel, Martinus Nijhoff, Dr W. Junk and MTP Press.

Sold and distributed in the U.S.A. and Canada
by Kluwer Academic Publishers,
101 Philip Drive, Norwell, MA 02061, U.S.A.

In all other countries, sold and distributed
by Kluwer Academic Publishers Group,
P.O. Box 322, 3300 AH Dordrecht, The Netherlands.

PREFACE

Rhizobium species involved in root nodule formation on
legume plants are one of the best known groups of microorganisms. The Rhizobium legume symbiosis continues to be of
strategic importance particularly in the context of food
production. As the world population grows, it is also necessary to have new developments taking place in crop improvement. The development and application of new technologies in
biological sciences over the past number of years have made
the entire area of plant-microbial interaction an exciting and
challenging research area to be involved in. In view of the
importance of symbiotic nitrogen fixation, it is not surprising that it still represents one of the priority areas for
commercial development in agricultural biotechnology. Since
this symbiosis involves an association between procaryotic and
eucaryotic partners, it requires of necessity a co-ordinated
and interdisciplinary approach. It was in this spirit that
this international conference was organised.

The scientific programme was designed to focus on physiological limitations affecting symbiotic nitrogen fixation and
the potential for overcoming such limitations using genetic
technologies. Participants were drawn from contractants of
the EEC DGVI "Energy in Agriculture" nitrogen fixation programme. The scientific programme was also supplemented with
invited scientists from Europe and North America to provide
appropriate expertise on the various conference topics. The
scientific programme, divided into three sessions of formal
lectures and three round-table discussion sessions, attracted
the interest of participating scientists from universities,
research centres and industry, concentrating on both funda-
mental research and its potential applications.

The conference would not have been possible without the
support of the main sponsors as well as the assistance from
the host institution, University College Cork. We are
indebted to our colleagues and to a number of other people
whose contributions helped make this conference a successful
event.

On a sadder note, just after the conference, it was with
great regret that we learned of the uxexpected death of our
distinguished colleague, Dr. Barry Chelm. To honour his many
contributions to nitrogen fixation research, we decided, in
association with his family and M. Kahn, to dedicate this
volume to his memory.

Dr. Fergal O'Gara,
December, 1987.

BARRY CHELM (1952-87)

Barry Chelm died of respiratory failure on Sept. 2, 1987, while on a camping trip. His unexpected and premature death at age 35 deprives the field of symbiotic nitrogen fixation research of one of its most promising members. In our small society, Barry's quick mind and even quicker sense of humour will always be remembered. In the wider world, his scientific work is recognised as innovative and solid, and his contributions will undoubtedly be recognized for some time to come.

For his Ph.D. degree with Richard Hallick at the University of Colorado, Barry studied RNA metabolism during chloroplast development in <u>Euglena</u> <u>gracilis</u>. Postdoctoral work with Peter Geiduschek at the University of San Diego focused on the analysis of RNA polymerase and promoter interactions in <u>Bacillus</u> <u>subtilis</u>. Among the several novel aspects of this work was the early use of gel electrophoresis to analyze initiation complexes. In 1981, Barry moved to Michigan Stat University and, as a member of the Plant Growth Laboratory and Department of Microbiology, began to investigate the genetics and physiology of <u>Bradyrhizobium</u> <u>japonicum</u>. From his laboratory have come a number of important and stimulating results. These include the discovery that the unique glutamine synthetase II enzyme of <u>B</u>. <u>japonicum</u> is similar to plant glutamine synthetases and that a bacterial mutant unable to make heme can still form an effective symbiosis.

No academic history can convey the essence of Barry as a colleague and friend. Barry was enthusiastic about ideas and about people and he was able to communicate his enthusiasm. He was a quick and precise critic with an unusual ability to uncover the weaknesses in an experiment or model without at the same time shaking the confidence of its author. Barry loved to talk and he loved to listen. Because his interests were broad, he could almost always find some common ground in science or elsewhere. We grew from his friendship and we will miss him.

T A B L E   O F   C O N T E N T S

SECTION III : GENETICS OF NITROGEN FIXATION

SECTION I : PHYSIOLOGICAL FACTORS AFFECTING NITROGEN FIXATION

THE INFLUENCE OF HOST PLANT ENERGY SUPPLY ON NITROGEN FIXATION

G. J. A. Ryle

AFRC Institute for Grassland and Animal Production,
Hurley, Maidenhead, Berkshire SL6 5LR.

ABSTRACT

In white clover plants dependent on N_2 fixation and acclimated to a uniform, artificial environment, N_2 fixation potential in the photoperiod is closely linked to current photosynthesis, while in the night it is limited by assimilate generated in the previous photoperiod. Darkening or defoliating plants reduces N_2 fixation potential in a manner which is quantitatively linked to the diurnal supply of assimilate. The effects of such treatments are apparent within 30–60 min. Enhancing photosynthesis above the level to which the plants are acclimated increases N_2 fixation potential, but the effect is relatively small.

INTRODUCTION

Rapid biological nitrogen fixation in legumes requires a substantial supply of respiratory substrates, provided directly or indirectly by photosynthesis, to provide the energy for the metabolic train associated with nitrogenase activity in the root nodules. In the forage legumes, where leaves and shoots provide the harvestable product the flux of photosynthetic products for growth and N_2 fixation is disrupted, to a greater or lesser extent, depending on the intensity and frequency of defoliation. The accurate measurement of the effects of defoliation on day-to-day rates of N_2 fixation presents insurmountable difficulties in the field. In such venues, measurements are laborious, expensive and provide only a general account of long-term N_2 fixation trends.

An earlier paper provided an account of some short-term changes in N_2 fixation potential in white clover when photosynthesis was disrupted by darkening and defoliation in controlled environment experimentation (Ryle et al., 1985). This paper extends these observations and provides a more detailed account of the relationships between nitrogenase activity in root nodules and the photosynthetic activity of shoot tissue.

MATERIALS AND TREATMENTS

Plants of white clover (<u>Trifolium</u> <u>repens</u> L. cv. Blanca), cultured

3

F. O'Gara et al. (eds.), Physiological Limitations and the Genetic Improvement of Symbiotic Nitrogen Fixation, 3–10.
© 1988 by Kluwer Academic Publishers.

from stolon tips, were grown singly in 1.3 dm^3 pots of Perlite in growth cabinets with a 12 h photoperiod, at a light intensity of 600 ± 20 μmol m^{-2} s^{-2} (Lamda PAR meter). The plants received a complete nutrient solution, added daily, containing all essential nutrients with the exception of nitrogen. Each pot was inoculated with 50 cm^3 of solution containing Rothamsted Rhizobium 221, as soon as the stolon tips developed roots. Measurements of root plus nodule respiration, acetylene reduction and other associated measurements were made on plants gently 'flooded' from their pots of Perlite and sealed into 1.2 dm^3 root chambers containing 10% nutrient solution (Fig. 1), after which they were returned to the growth cabinet with air bubbling through the root chamber until their respiration effluxes had equilibrated. For experimental measurements, pairs of plants sealed in their root chambers were enclosed within a ventilated assimilation chamber, the whole contained within a growth cabinet which provided light and temperature control for the aerial parts of the two plants. The photosynthesis of the shoots and the respiration of nodulated roots was measured separately by infra-red gas analysis, and the gas mixture passing over shoots and roots varied according to need.

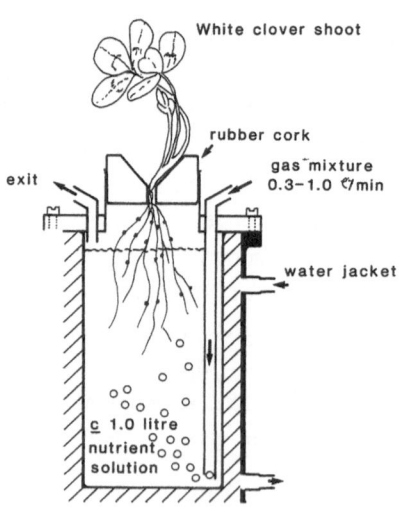

Fig. 1   Root assimilation chamber.

RESULTS AND DISCUSSION

The general relationships between the respiration of roots and nodules, and between the respiration of nodules and their ability to reduce acetylene, have been reported for the Blanca/221 symbiosis (Ryle et al., 1985; Ryle et al., 1986). The equilibrated respiration of the nodulated root system of white clover includes a component associated with root growth and maintenance (including ion uptake) and a much larger component linked to the fixation of N_2 and other associated metabolic processes in the root nodules. Conventionally, potential N_2 fixation is measured by exposure of nodulated root systems to a gas mixture containing 10% acetylene and measuring the rate of ethylene production. In our experiments, mature rapidly-growing white clover plants exhibit a rapid peak of C_2H_4 production after a few minutes exposure to 10% C_2H_4, followed by a decline and plateau, as shown in Fig. 2 (cf. Minchin et al., 1983). Root plus nodule respiration reflects this peak and plateau response. Sheehy et al., (1983) interpret this response in terms of a change in oxygen diffusion resistance. Furthermore, if the level of oxygen in the gas mixture is reduced from ambient levels, both C_2H_4 and respiration production rates also progressively fall in the Blanca/221 symbiosis (Ryle et al., 1986), as they do in other symbioses. Fig. 2 demonstrates the effect of decrease of oxygen level from 21% to 3% in a single step. Other comparisons of the responses of nodulated root systems to low levels of oxygen, and to physical removal of nodules, indicate that a brief exposure to 3% O_2 effectively halts all or nearly all nitrogenase-linked respiration with little immediate effect on root respiration or basal respiration in the nodule. Continuous measurements of nodulated root respiration plus brief measurements of respiration in 3% O_2 thus provide, by difference, an account of that nodule respiration linked to nitrogenase activity. In turn, it has been shown that such nitrogenase-linked respiration is linearly related to acetylene reduction activity in both defoliated and non-defoliated white clover plants up to a maximum of 80 μmol C_2H_4 h^{-1} plant^{-1} (c. 25 μmol N_2 h^{-1} plant^{-1}) (Ryle et al., 1986). While 'nitrogenase-linked respiration' (NLR) provides a simple, effective measure of N_2 fixation potential, it is not as accurate as a direct acetylene reduction assay.

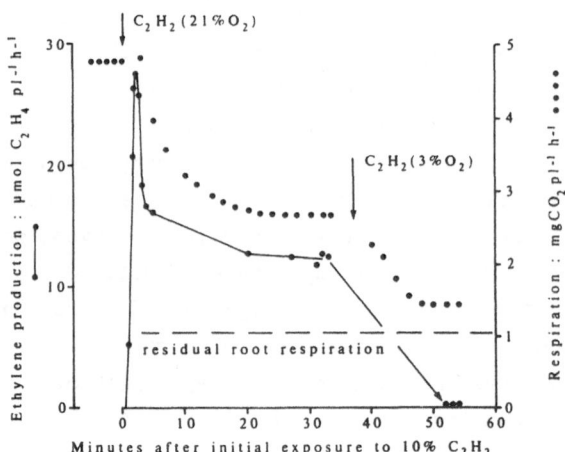

Fig. 2   Respiration and acetylene reduction in white clover
in 21% and 3% oxygen.

Direct, continuous measurements of respiration have shown that
white clover plants wholly dependent on N_2 fixation for nitrogen, and
acclimated to favourable growing conditions in controlled
environments, exhibit a rapid decline in NLR if they are darkened
early in the photoperiod, if they are defoliated, or if their
photosynthesis is inhibited by DCMU (Ryle et al., 1985).

The effect of darkening during a normal 12 h photoperiod was
further investigated by darkening pairs of plants at 12 progressively
later times during the photoperiod. In all plants, darkening affected
NLR during the current diurnal (day plus night) period. The effect
was most marked when the plants had recently emerged from a 12 h night
and decreased in a sigmoid manner as darkening was delayed during the
photoperiod (Fig. 3).

The effect of time of complete defoliation (removal of all
expanded leaves, but leaving all other tissue intact) during a 12 h
photoperiod was generally small. Generally, NLR declined within a few
hours to values close to zero after complete defoliation. Where such
defoliation was delayed until late in the photoperiod there was a
small but perceptible extension of the period before NLR began to
decline; subsequently the rate of decline was little slowed. When
plants were partially defoliated, NLR also declined but the rate of

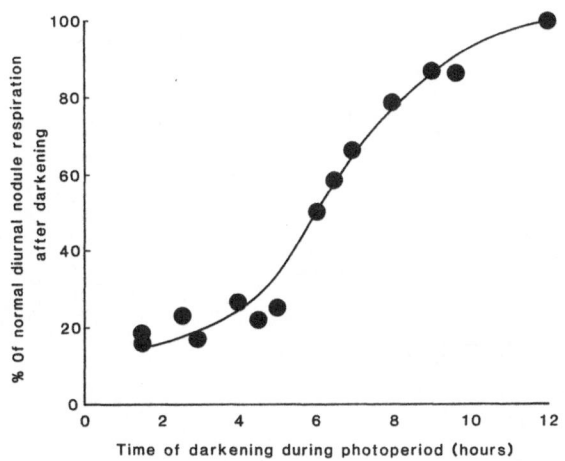

Fig. 3   Time of darkening and NLR in white clover.

decline was slower, exhibited some recovery during the photoperiod and
maintained a higher rate during the night period (Fig. 4).   The
short-term oscillations in NLR illustrated in Fig. 4 are
characteristic of partially defoliated plants;   they presumably
reflect adaptive responses to stress by the plant, perhaps involving
mobilisation of energy resources not normally utilized during the
fixed cycle of day and night in plants acclimated to controlled

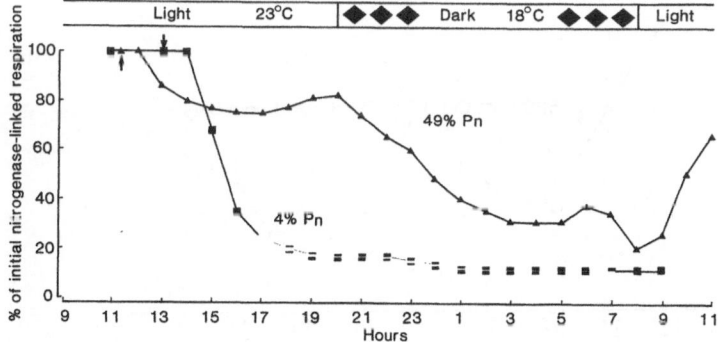

Fig. 4   Effect of complete (4% normal net photosynthesis)
and partial defoliation (49% normal net photosynthesis) on
NLR in white clover.

8

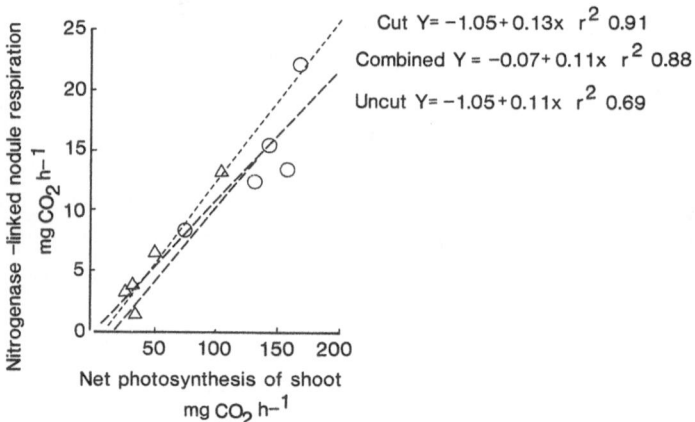

Cut $Y = -1.05 + 0.13x$ r^2 0.91

Combined $Y = -0.07 + 0.11x$ r^2 0.88

Uncut $Y = -1.05 + 0.11x$ r^2 0.69

Fig. 5   Shoot photosynthesis and NLR in defoliated (triangles) and undefoliated (circles) white clover.

environments. A generalized relationship between NLR and net photosynthesis in such plants acclimated to 23/18°C day and night temperatures, before and after defoliation, is illustrated in Fig. 5. In the short-term, NLR is clearly linearly related to current supply of photosynthate.

The treatments discussed thus far all reduce the supply of

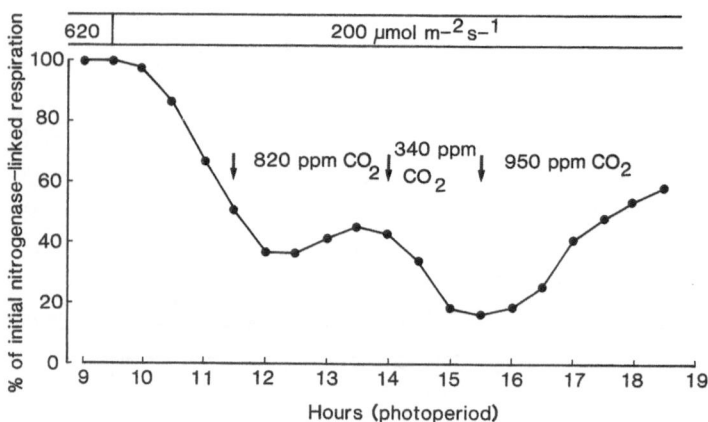

Fig. 6   Effect of enhanced shoot photosynthesis on NLR in a reduced light intensity.

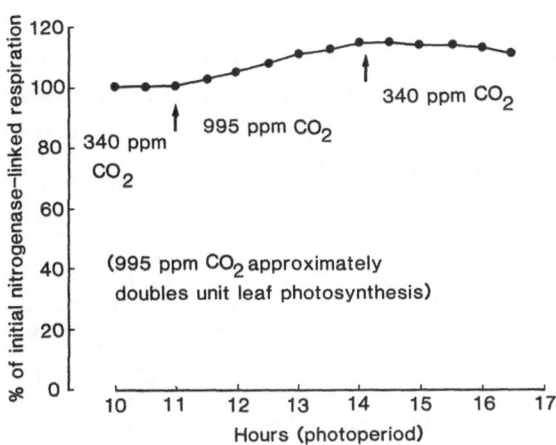

Fig. 7   Effect of enhanced shoot photosynthesis on NLR in full light.

assimilate for growth and N_2 fixation. Figs. 6 and 7 illustrate examples where elevated CO_2 levels were used to increase photosynthesis. In general, CO_2 levels of 950–1000 ppm approximately double the normal rate of leaf photosynthesis. In Fig. 6, enhanced photosynthesis partially offsets the fall in NLR resulting from a three-fold reduction in the normal light intensity. In Fig. 7, approximately doubling rate of photosynthesis enhanced NLR in normal bright illumination by about 15% in 3 hours. Subsequently, with normal photosynthesis, NLR declined.

In these white clover plants growing rapidly, dependent on N_2 fixation and acclimated to a fixed 12 h controlled-environment light-dark cycle, supply and utilization of assimilate are clearly finely balanced and any disruption of photosynthesis is manifested in a decline in NLR. Where the disruption was due to partial or complete darkening, the effect decreased the later it was imposed, suggesting that plants could mobilise assimilate stored earlier in the same photoperiod to replace that provided by current photosynthesis, at least until it was consumed. The rapid fall in NLR in defoliated plants, where some net photosynthesis still persisted, imples that assimilate mobilisation from sites outside the leaves was either not possible or insufficient to meet the needs of N_2 fixation. The experiments with enhanced CO_2 levels indicate that, in the short-term, more assimilate was not in itself sufficient to markedly increase NLR in plants growing in an environment to which they were acclimated, although it did so when the normal level of photosynthesis had already been depressed. The implication is that a nodule enzyme system had been developed sufficient in size and/or efficiency to utilize the assimilate available in that environment.

REFERENCES

Minchin, F.R., Witty, J.F., Sheehy, J.E. and Muller, M. 1983. A major error in the acetylene reduction assay: decreases in nodular nitrogenase activity under assay conditions. J. Exp. Bot., 34, 641-9

Ryle, G.J.A., Powell, C.E. and Gordon, A.J. 1985. Short-term changes in CO_2 evolution associated with nitrogenase activity in white clover in response to defoliation and photosynthesis. J. Exp Bot. 36, 634-643.

Ryle, G.J.A., Powell, C.E. and Gordon, A.J. 1986. Defoliation in white clover: nodule metabolism, nodule growth and maintenance, and nitrogenase functioning during growth and regrowth. Ann. Bot. 57, 263-271.

Sheehy, J.E., Minchin, F.R. and Witty, J.F. 1983. Biological control of the resistance to oxygen flux in nodules. Ann. Bot. 52, 565-71.

CARBON METABOLISM AND THE EXCHANGE OF METABOLITES
BETWEEN SYMBIONTS IN LEGUME NODULES

J.G. Streeter and S.O. Salminen
The Ohio State University
Ohio Agricultural Research and Development Center
Department of Agronomy
Wooster, OH 44691-6900, USA

ABSTRACT

Two examples of exchange of carbon between host and bacteroids are presented. In the first example, α,α-trehalose which accumulates in nodules is synthesized in bacteroids, but substantial quantities of trehalose are released to the host cytoplasm where it is "recycled" to glucose. In comparisons across *Bradyrhizobium japonicum* strains, trehalose concentrations in nodules are negatively correlated with acetylene reduction activity. We still have no explanations for these unusual relationships involving trehalose. In the second example we have preliminary evidence for the operation of a malate/aspartate shuttle in *B. japonicum* bacteroids. Malate is taken up and, after oxidation and transamination, is returned to medium as aspartate. Glutamate which may be taken up or synthesized in bacteroids, is transaminated to α-ketoglutarate, which is exported to the medium. Evidence from the literature and from our recent labeling experiments supports this model, but it is far from being established.
We suggest that a concept of metabolite exchange between symbionts is more appropriate than a unidirectional flow of reduced carbon into bacteroids and reduced nitrogen out of bacteroids.

INTRODUCTION

Current concepts of carbon metabolism in legume nodules begin with sucrose which is clearly the major source of reduced carbon entering all legume nodules which have been studied. Enzymes of glycolysis, which are very active in host cytoplasm, convert sugars to organic acids. Organic acids, malate and succinate in particular, are thought to pass through the peribacteroid membrane and serve as the main sources of reduced carbon for bacteroids. It is generally indicated that acids are oxidized to CO_2 by bacteroids with the concomitant "extraction" of electrons which, in turn, are used for vital bacteroid functions including the operation of nitrogenase.

The above concepts are supported by numerous lines of evidence, including the following:

• Organic acids are rapidly absorbed by an active uptake mechanism, whereas carbohydrates are only slowly absorbed by bacteroids (Reibach and Streeter, 1984).

11

F. O'Gara et al. (eds.), *Physiological Limitations and the Genetic Improvement of Symbiotic Nitrogen Fixation, 11–20.*
© 1988 by Kluwer Academic Publishers.

• Organic acids support higher rates of respiration and nitrogenase activity by isolated bacteroids than are supported by sugars (Saroso, et al., 1984).

• Rhizobia which lack the ability to absorb dicarboxylic acids form Fix⁻ nodules. This has been observed independently by several different groups (Ronson, et al., 1981; Finan, et al., 1983; Arwas, et al., 1985).

• Bacteroids have high activity of enzymes for organic acid metabolism, but low activity of key enzymes of glycolysis, the pentose phosphate and Entner-Doudoroff pathways (Saroso, et al., 1986; Salminen and Streeter, 1986b).

Although the concepts described in the first paragraph are widely accepted, our thesis here is that a concept of carbon exchange between symbionts should be considered. Two examples of carbon exchange are described. The first, involving α,α-trehalose is well documented, but we presently have no knowledge of its importance. The second involves the possible operation of a malate/aspartate shuttle in nodules. Our evidence for this is very preliminary, but if correct, the concept could be highly important in our attempts to understand nodule function.

TREHALOSE

This glucose-glucose disaccharide is the blood sugar of insects and is probably present in all fungi, the organisms where it has been most thoroughly studied. Recently, nitrogen fixing organisms including cyanobacteria, actinomycetes, and *Rhizobium* have been found to accumulate trehalose (Streeter, 1985). The sugar is not synthesized in higher plants where, in fact, it may be toxic unless the hydrolytic enzyme trehalase is present.

Trehalose accumulates in all legume nodules we have examined and the concentration present depends on the *Rhizobium* strain. In soybean (*Glycine max*) nodules formed by indigenous *B. japonicum* and harvested weekly, the seasonal mean (72 observations) concentration of nodule sugars included sucrose (2.8 mg/g fresh wt of nodule), glucose (1.4), trehalose (1.3), maltose (0.41), and fructose (0.31). Several lines of evidence have proven that trehalose in nodules is synthesized in bacteroids; for example, the key enzyme of trehalose synthesis could be found only in bacteroids (Salminen and Streeter, 1986a). Trehalose comprised 65% to as

little as 15% of the mono- and disaccharides in bacteroids depending on
Rhizobium strain (Streeter, 1985).

Two recent findings regarding trehalose are surprising and puzzling:

• Bacteroids may contain as little as 20% of the trehalose in the
nodule, depending on *Rhizobium* strain. (This low proportion is not due to
loss of trehalose from bacteroids during the isolation procedure
(Streeter, 1985).) Most of the trehalase activity is in the host
cytoplasm but we do not know how accessible the enzyme is to trehalose
released from bacteroids. The proportion of trehalose retained by
bacteroids tends to increase with increasing nodule age, but an average
across strains and nodule ages indicates that about 50% of the trehalose
in a soybean nodules is not located in the bacteroids where it was
synthesized (Streeter, 1985). Although we have not attempted to measure
the flux of carbon through this apparently "futile cycle" (trehalose
synthesis in bacteroids and trehalose hydrolysis in the host cytoplasm),
it seems highly unusual for what appears to be substantial quantity of
reduced carbon to be released by bacteroids.

• In a comparison of *B. japonicum* strains a negative correlation
between acetylene reduction activity (nodulated roots) and trehalose
concentration in nodules was found (Fig. 1). Concentrations of other
sugars were not statistically related to acetylene reduction activity.
The relationship with trehalose concentration has been noted in two
different experiments involving four different nodule ages; thus, it seems
unlikely that the relationship is a spurious one.

Rhizobium strains which accumulate high trehalose in culture also
tend to form nodules with high trehalose concentration. Thus, for the
results in Fig. 1 we can suggest that bacteroids which synthesize more
trehalose tend to have lower nitrogenase activity. Perhaps additional
trehalose is synthesized merely because there is less demand for reduced
carbon in these bacteroids. We do not think that trehalose serves as an
important source of reductant for nitrogenase; this role is not consistent
with major release of the sugar from bacteroids. At the present time we
are searching for mutants of *Rhizobium* which cannot synthesize or cannot
break down trehalose. We expect that these mutants will help us to
determine why so much trehalose is released from bacteroids and why
trehalose concentration is associated (negatively) with nodule nitrogenase
activity.

14

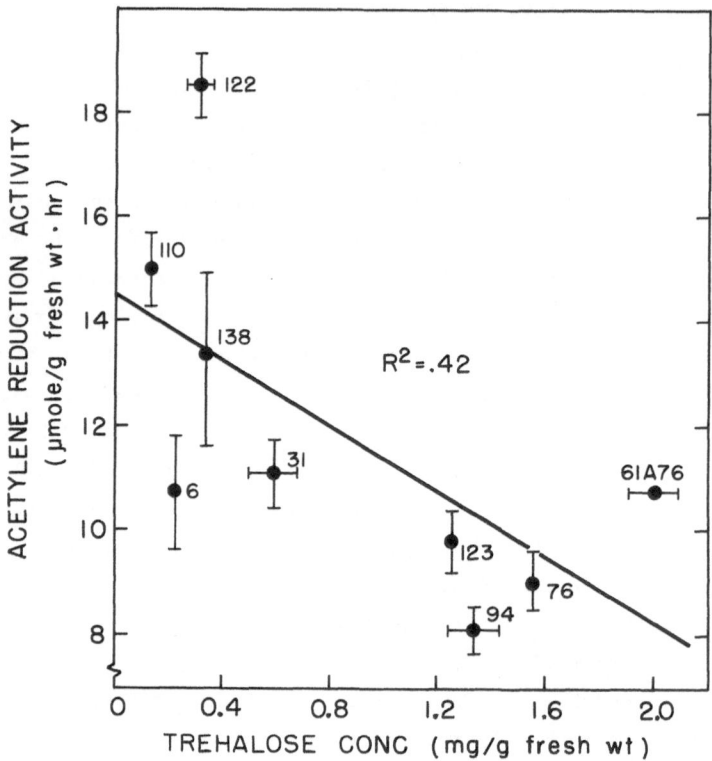

Fig. 1   Relationship between acetylene reduction activity and
trehalose concentration in soybean nodules. Vertical and
horizontal bars are standard errors of 4 replicates. Numbers
indicate USDA strains except for 61A76 (Nitragin Co.).

MALATE/ASPARTATE SHUTTLE

In recent studies of the metabolism of dicarboxylic acids by
anaerobically isolated B. *japonicum* bacteroids, the analysis of labeled
compounds in the cells was emphasized. We were surprised to find that
nearly 50% of ^{14}C-succinate taken up was recovered in the amino acid
fraction and that nearly all of this amino acid label was in a single
compound, namely glutamate (Salminen and Streeter, 1987). We compared
[14]C-labeled malate, glutamate, and aspartate with succinate and obtained a
similar result; namely, the compound which was, by far, the most heavily
labeled in bacteroids was glutamate. (All compounds were, of course,
partially converted to [14]CO_2, malate being the most rapidly and aspartate
the least rapidly converted.) The ^{14}C-labeling of glutamate is consistent

with the previously observed labeling of bacteroid glutamate by $^{15}N_2$ (Ohyama and Kumazawa, 1980). We concluded that there is a "substantial" pool of glutamate in bacteroids (Salminen and Streeter, 1987), and have subsequently found that the concentration of glutamate in bacteroids is high relative to other amino acids and organic acids (Streeter, in press).

What might be the reason for this large glutamate pool which is rapidly labeled from ^{14}C-substrates or from $^{15}N_2$? We considered many possible answers to this question and have recently adopted as a working hypothesis the operation of a malate/aspartate shuttle in bacteroids. The shuttle provides a mechanism for the transfer of reducing equivalents from one compartment to a second compartment because malate is oxidized on one side of the barrier and reformed on the other side (Fig. 2). The shuttle requires the same two enzymes on both sides of the barrier, namely malate dehydrogenase and aspartate aminotransferase (also called

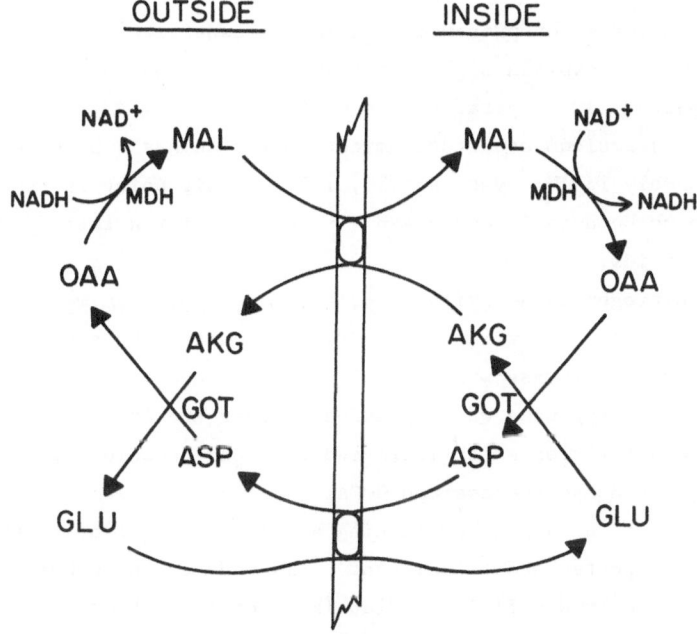

Fig. 2   The malate/aspartate shuttle, a mechanism for the transfer of reducing equivalents in which no net transfer of carbon occurs. Abbreviations: MAL = malate, OAA = oxalacetate, ASP = aspartate, GLU = glutamate, AKG = α-ketoglutarate; MDH = malate dehydrogenase, GOT = glutamate:oxalacetate transaminase (also named aspartate amino-transferase).

glutamate:oxalacetate transaminase). The shuttle also requires two co-transport activities in the barrier (membrane), and neither of these transporters is thought to require energy for its operation. There is some evidence for the operation of a malate/aspartate shuttle in *Alnus glutinosa* nodules (Akkermans, et al., 1981), and the possibility of a shuttle has also been suggested previously for legume nodules (Kahn, et al., 1985).

As we examined the literature, we found a surprisingly large amount of support for the shuttle hypothesis:

• High activity of malate dehydrogenase (Waters, et al., 1985) and aspartate aminotransferase (Ryan, et al., 1972; Reynolds, et al., 1981) are found in both cytosol and bacteroids of legume nodules. Malate dehydrogenase would, of course, be required for functions other than the shuttle. Cytosolic aminotransferase might be required for aspartate and asparagine synthesis; thus, these high enzyme activities do not necessarily indicate the presence of a malate/aspartate shuttle. However, it is difficult to explain high aminotransferase in bacteroids on the basis of aspartate and asparagine synthesis.

• The *B. japonicum* aspartate aminotransferase has a high Km for glutamate, namely 13 mM (Ryan, et al., 1972). This point is especially important to us because it could explain the need for a large glutamate pool in bacteroids.

• When nitrogenase activity in cultured *B. japonicum* was induced by lowering $[O_2]$, etc., several enzyme activities also increased including aspartate aminotransferase which nearly doubled in activity relative to cells in which nitrogenase was repressed (Werner and Stripf, 1978). (Other enzyme activities which increased upon derepression of nitrogenase included alanine dehydrogenase and GOGAT.)

• A mutant of *B. japonicum* lacking aspartate aminotransferase has recently been reported and this mutant forms nodules on soybeans, but the plants are extremely N deficient (Zlotnikov, et al., 1984). What is very curious is that nodules were reported to have acetylene reduction activity. Thus, this mutant needs to be more completely characterized before conclusions can be drawn.

We have recently begun to test the shuttle hypothesis by extending our labeling studies with anaerobically isolated bacteroids to include the use of multiple substrates and the analysis of labeling patterns in the reaction mixture. Note (Fig. 2) that, if the shuttle is operating, one might expect to find some labeling of aspartate in the reaction mixture when ^{14}C-malate is provided and especially if glutamate is also provided to facilitate the export of aspartate. We did, in fact, find that adding unlabeled glutamate stimulated the labeling of aspartate in the reaction mixture relative to no addition or to adding alanine (Table 1). The most dramatic effect was, however, provided by the aspartate treatment which gave about a 16-fold increase in labeling of reaction mixture aspartate. We attribute this to a "trapping" effect and suggest that ^{14}C-aspartate which is exported from bacteroids is diluted out (trapped) by the unlabeled compound. Alternatively, the presence of unlabeled aspartate may stimulate the release of ^{14}C-aspartate by some mechanism which we do not understand.

TABLE 1   Effect of adding a second substrate on the labeling of amino acids in the reaction mixture from ^{14}C-malate supplied to *B. japonicum* bacteroids.[a]

| Second Substrate Added | Radioactivity (total dpm x 10^{-3}/sample) | | |
	Glutamate	Aspartate	Alanine
None	1.8	2.6	8.2
Glutamate	5.7	5.7	6.2
Aspartate	5.6	41.8	8.7
Alanine	2.7	1.0	9.6

[a] Bacteroids were isolated anaerobically and incubated with myoglobin plus 3% O_2 in the gas phase. The concentration of each substrate was 0.5 mM and about 2 x 10^6 dpm of ^{14}C-malate were supplied. Data are averages from 15 and 30 min incubations.
In bacteroids, labeling (dpm x 10^{-3}) of the three amino acids was as follows: glutamate (176), aspartate (2.6), and alanine (33). There were no significant effects of the second substrate on the labeling of these three compounds in bacteroids.

Whatever the mechanism, a similar result was obtained in the analogous experiment with ^{14}C-glutamate. The model (Fig. 2) would predict the labeling of α-ketoglutarate in the reaction mixture. Alpha-ketoglutarate was labeled only when a second substrate was provided and,

in what we think is a significant result, malate was much more active than
succinate in promoting the labeling of AKG (Table 2). When AKG itself was
provided as the second substrate, there was massive labeling of AKG in the
reaction mixture. Other "second substrate effects" were more complicated
than in the experiment with ^{14}C-malate and we do not really understand
these (Table 2).

TABLE 2   Effect of adding a second substrate on the labeling
of organic acids in the reaction mixture from ^{14}C-glutamate
supplied to B. *japonicum* bacteroids.[a]

Second Substrate Added	Radioactivity (total dpm x 10^{-3}/sample)		
	Malate	Succinate	α-Ketoglutarate
None	0.2	0.3	0
Malate	138.0	50.0	84.5
Succinate	9.7	9.9	15.0
α-Ketoglutarate	3.9	43.4	601.0

[a] Conditions as in Table 1 except that data are averages for 30
and 60 min incubations. In bacteroids, succinate and malate
contained <5 x 10³ dpm on the average and labeling was not
influenced by the second substrate. Labeling of α-ketoglutarate in bacteroids was too low to quantify. Labeling of
other keto acids in the reaction mixture was essentially nil.

Our experiments to date are very preliminary and raise more questions
than they answer. One question of crucial importance is, if one assumes
that a shuttle is operative, are transport activities in the peribacteroid
membrane, in bacteroid membranes, or both? We do not know the answer.
However, we think that the second substrate experiments may at least
indicate that bacteroids could support a metabolite shuttle even if they
may not control transport. These experiments also support a second
general conclusion: that bacteroids appear to be well suited to
participate in a very active exchange of metabolites when supplied with
multiple substrates.

One final point. By emphasizing the malate/aspartate shuttle we do
not mean to suggest that this is likely to be the only mechanism by which
bacteroids obtain reducing equivalents. The literature indicates that

multiple mechanisms may be involved. For example, work on dicarboxylate uptake mutants, mentioned earlier, indicates that dicarboxylic acid uptake, per se, is required for N_2 fixation in bacteroids. Perhaps a shuttle of some sort operates in conjunction with dicarboxylate uptake.

CONCLUSIONS

The concept of unidirectional flows of reduced carbon into and reduced nitrogen out of bacteroids probably represents an over-simplification. We know that trehalose is synthesized in bacteroids and that much of the sugar is released to the cytosol. We think that a malate/aspartate shuttle may operate in legume nodules as one mechanism for the transfer of reducing equivalents to bacteroids. The possibility that multiple substrates are provided by the host to bacteroids should be explored further. Finally, we would urge workers interested in metabolism relating to symbiotic nitrogen fixation to be open to the idea that an exchange of metabolites between symbionts may be the most appropriate model for the operation of nodules.

REFERENCES

Akkermans, A.D.L., Huss-Danell, K. and Roelofsen, W. 1981. Enzymes of the tricarboxylic acid cycle and the malate-aspartate shuttle in the N_2-fixing endophyte of *Alnus glutinosa*. Physiol. Plant., 53, 289-294.

Arwas, R., McKay, I.A., Rowney, F.R.P., Dilworth, M.J. and Glenn, A.R. 1985. Properties of organic acid utilization mutants of *Rhizobium leguminosarum* strain 300. J. Gen. Microbiol., 131, 2059-2066.

Finan, T.M., Wood, J.M. and Jordan, D.C. 1983. Symbiotic properties of C_4-dicarboxylic acid transport mutants of *Rhizobium leguminosarum*. J. Bacteriol., 154, 1403-1413.

Kahn, M.L., Kraus, J. and Somerville, J.E. 1985. A model of nutrient exchange in the *Rhizobium*-legume symbiosis. In "Nitrogen Fixation Research Progress" (Eds. H.J. Evans, P.J. Bottomley and W.J. Newton). (Martinus Nijhoff, Dordrecht), pp. 193-199.

Ohyama, T. and Kumazawa, K. 1980. Nitrogen assimilation in soybean nodules. II. $^{15}N_2$ assimilation in bacteroid and cytosol fractions of soybean nodules. Soil Sci. Plant Nutr., 26, 205-213.

Reibach, P.H. and Streeter, J.G. 1984. Evaluation of active versus passive uptake of metabolites by *Rhizobium japonicum* bacteroids. J. Bacteriol., 159:47-52.

Reynolds, P.H.S., Boland, M.J. and Farnden, K.J.F. 1981. Enzymes of nitrogen metabolism in legume nodules: Partial purification and properties of the aspartate aminotransferases from lupine nodules. Arch. Biochem. Biophys., 209, 524-533.

Ronson, C.W., Lyttleton, P. and Robertson, J.G. 1981. C_4-dicarboxylate transport mutants of *Rhizobium trifolii* form ineffective nodules on *Trifolium repens*. Proc. Natl. Acad. Sci. USA, 78, 4284-4288.

20

Ryan, E., Bodley, F. and Fottrell, P.F. 1972. Purification and characterization of aspartate aminotrasferases from soybean root nodules and *Rhizobium japonicum*. Phytochem., 11, 957-963.

Salminen, S.O. and Streeter, J.G. 1986a. Enzymes of α,α-trehalose metabolism in soybean nodules. Plant Physiol., 81, 538-541.

Salminen, S.O. and Streeter, J.G. 1986b. Uptake and metabolism of carbohydrates by *Bradyrhizobium japonicum* bacteroids. Plant Physiol., 83, 535-540.

Salminen, S.O. and Streeter, J.G. 1987. Involvement of glutamate in the respiratory metabolism of *Bradyrhizobium japonicum* bacteroids. J. Bacteriol., 169, 495-499.

Saroso, S., Dilworth, M.J. and Glenn, A.R. 1986. The use of activities of carbon catabolic enzymes as a probe for the carbon nutrition of snakebean nodule bacteroids. J. Gen. Microbiol., 132, 243-249.

Saroso, S., Glenn, A.R. and Dilworth, M.J. 1984. Carbon utilization by free-living and bacteroid forms of cowpea *Rhizobium* strain NGR234. J. Gen. Microbiol., 130, 1809-1814.

Streeter, J.G. 1985. Accumulation of α,α-trehalose by *Rhizobium* bacteria and bacteroids. J. Bacteriol., 164, 78-84.

Waters, J.K., Karr, D.B. and Emerich, D.W. 1985. Malate dehydrogenase from *Rhizobium japonicum* 3I1b-143 bacteroids and *Glycine max* root nodule mitochondria. Biochem., 24, 6479-6486.

Werner, D. and Stripf, R. 1978. Differentiation of *Rhizobium japonicum*, I. Enzymatic comparison of nitrogenase repressed and derepressed free-living cells and of bacteroids. Z. Naturforsch., 33c, 245-252.

Zlotnikov, K.M., Marunov, S.K. and Khmel'nitskii, M.I. 1984. Disturbance in assimilation of fixed nitrogen by soybean plants in symbiosis with the ASP⁻ bacterium *Rhizobium japonicum*. Dokl. Akad. Nauk SSSR, 275, 189-192.

BACTERIAL CATABOLISM OF NITROGEN CONTAINING COMPOUNDS IN SYMBIOTIC NITROGEN FIXATION

M. L. Kahn*, J. Kraus**, R. G. Shatters**

*Department of Microbiology and Institute of Biological Chemistry
**Program in Genetics and Cell Biology
Washington State University
Pullman, Washington, USA 99164

ABSTRACT

Recent work indicating that there may be a flow of fixed nitrogen to Rhizobium bacteroids is discussed in light of a hypothesis that such a flow may be important to nodule function. Evidence that nitrogen is transported to the microsymbiont includes analysis of auxotrophic mutants, consideration of proline metabolism in nodules, and the finding of unusual amino compounds within nodules that can specifically support bacteria growth. We describe experiments that implicate an NAD dependent glutamate dehydrogenase in glutamate catabolism by free living bacteria.

INTRODUCTION

In the symbiosis between leguminous plants and Rhizobium or Bradyrhizobium bacteria, the plant provides the bacteria with carbon and energy and in return it obtains ammonia produced by bacterial nitrogen fixation. Which specific compounds are provided to the bacteria has been a major question of research since it has generally been felt that one method of increasing yields due to symbiotic nitrogen fixation would be to increase the amount of energy available to the bacteria. This depends both on knowing how chemical energy is transferred to the bacteroid and how the bacteroid uses this energy to fix nitrogen.

The carbon compounds delivered to the nodule are primarily monosaccharides and disaccharides, chiefly glucose and sucrose (Emerich et al., 1983). However, strong arguments have been presented that these compounds are not the primary carbon sources for the bacteroids (Dilworth and Glenn, 1985). These sugars are therefore likely to be converted by the plant into other compounds which are delivered to the bacteroid through the peribacteroid membrane.

Dicarboxylic acids are thought by many to be the class of compounds most likely to serve as primary carbon source (reviewed in Ronson and Astwood, 1985). Mutants of various bacterial species that are blocked in

21

F. O'Gara et al. (eds.), Physiological Limitations and the Genetic Improvement of Symbiotic Nitrogen Fixation, 21–27.
© 1988 by Kluwer Academic Publishers.

dicarboxylic acid transport are Fix-. In addition, succinate supports the highest levels of nitrogen fixation by isolated bacteroids and other dicarboxylic acids are also excellent substrates in this system.

Kahn et al. (1985) have proposed that a carbon compound that contains nitrogen might also be important as an energy or carbon source for the bacteroids. The model had two major components:

1. Bacteroids do not appear to be nitrogen stressed, they have low levels of nitrogen assimilating enzymes and export ammonia. If they are not nitrogen stressed, it is peculiar that the bacteroids continue to reduce nitrogen, a process that requires significant energy.

2. If the bacteroid's carbon supply was dependent on the plant's nitrogen status, there would be a reason for them to fix nitrogen despite the abundance of ammonia available. Kahn et al. proposed that, by coupling the metabolism of the symbionts through the exchange of compounds that contained nitrogen, a more evolutionary stable relationship between the plant and its symbiont would be produced.

Although the use of amino acids by the bacteroid had been considered previously, the general view seemed to be that the flow of nitrogen was unidirectional and away from the bacteroid. However, because of the reversibility of many of the enzymes used in amino acid metabolism, most of the data in the literature are consistent with either an anabolic or catabolic role for amino acids. In this paper some recent evidence concerning nitrogen flow and catabolism will be reviewed.

DOES NITROGEN FLOW TO THE BACTEROID?

Two sorts of data have previously supported the view that at least some nitrogen containing compounds are available to the endosymbiont. The first is that root exudates contain relatively high concentrations of amino acids, especially glutamate. Although it can be argued that compartmentalization within the nodule prevents these compounds from reaching the bacteroids, the additional observation that amino acid auxotrophs are generally Fix+ suggests that sufficient quantities of amino acids are available within the infected cells.

In R. meliloti, this argument is especially strong with respect to the amino acid glutamate. Because glutamate and glutamine are used as amino group donors in a number of biosynthetic reactions and glutamate is thought to be an important counteranion in at least some bacteria (Richey et al., 1987), the amount of these amino acids needed for growth is much larger than the average. A number of groups have isolated mutants that cannot assimilate glutamate because of defects in glutamate synthase (GOGAT) and these mutants are Fix+. We have recently constructed a mutant that lacks both glutamine synthetase isoforms I and II. The mutant is not an auxotroph in a strict sense since, although it cannot grow using either glutamate or ammonia as sole nitrogen source, it will grow slowly when given high concentrations of both of these compounds. This might be due to a third glutamine synthetase enzyme (Kumar and Rao, 1986) or to some activity of an enzyme, such as asparagine synthetase or carbamoylphosphate synthetase, that can use either ammonia or the amide nitrogen of glutamine to carry out its reaction. Although we do not yet understand why this mutant is leaky, we have been unable to grow it unless high concentrations of either glutamate or glutamine are present. Like the GOGAT mutants, the GSI⁻ GSII⁻ mutant is Fix+.

It must be emphasized that, although glutamate biosynthesis does not seem to be needed for effective symbiosis between R. meliloti and alfalfa, other symbiotic relationships are disrupted by glutamate auxotrophy. In the R. japonicum/soybean symbiosis, O°Gara et al. (1984) have found that GOGAT mutants are Fix-. Carlson et al. (1987) have recently constructed a GSI- GSII- mutant of R. japonicum and find that this mutant also produces Fix- nodules. R. sesbania mutants with defects in glutamate biosynthesis are Fix- (Donald and Ludwig, 1984). This difference from the situation with R. meliloti may be due to some leakiness of the R. meliloti auxotrophs when they differentiate into bacteroids or may reflect a basic difference between the R. meliloti/alfalfa symbiosis and these others. It is also possible that the Fix- phenotype is due to a problem establishing nitrogen fixation in nodules rather than in maintaining it since the catabolism of proline (see below) should produce considerable glutamate.

PROLINE METABOLISM.

Kohl et al. (1987) have recently presented evidence that proline may be catabolized by bacteroids of R. japonicum. Soybean nodules export much of their fixed nitrogen as ureides and since these are degradation products of purines, purine biosynthesis is very active in nodules. NADP+

is required for the synthesis of ribose-5-phosphate, the phosphorylated sugar used in making purines. In mammalian systems NADP+ is produced by pyrroline-5-carboxylase reductase (PCR), an enzyme involved in proline biosynthesis that shows a strong preference for NADP+ over NAD+. Kohl et al. find very high PCR activity in the plant cytosol fraction of soybean nodules, and suggest that considerable proline biosynthesis is occurring. Because proline is not exported from the nodule, they looked for the proline catabolic enzyme, proline dehydrogenase, and found a high activity in isolated bacteroids. They suggest that there is a cyclic flow of the carbon and nitrogen skeleton of proline between the plant and bacterial compartments with reduction of this skeleton to proline in the plant followed by oxidation in the bacteroid.

RHIZOPINES.

When Agrobacterium tumefaciens induces a tumor on a dichotyledonous plant, it transfers DNA into the plant cell that leads to abnormal growth of the cell and also programs the plant to produce specific unusual compounds that can be catabolized by the bacteria. These compounds are called opines and are generally formed by joining an amino acid to a sugar or organic acid. Recently, Murphy et al. (1987) have suggested that an analogous situation may exist in the interaction between R. meliloti L5-30 and alfalfa. They have shown that L-3-O-methyl-scyllo-inosamine (3-O-MSI) is contained in nodules formed by strain L-50, that the presence of this compound depends on genes carried on one of the R. meliloti megaplasmids and that the ability to catabolize this compound is also carried on the plasmid. They have not yet established which of the symbionts synthesizes 3-O-MSI or whether there is transfer of DNA from the bacteria to the plant. However, in the context of this paper it may be significant that opines contain nitrogen and probably release their amino acid or amine component during catabolism.

GLUTAMATE CATABOLISM IN RHIZOBIUM.

Salminen and Streeter (1987) have studied the metabolism of malate, succinate, aspartate and glutamate by isolated bacteroids of B. japonicum. They found that the organic acids were absorbed about twice as rapidly as the amino acids but that all were degraded fairly efficiently to CO_2.

A substantial fraction of all of the labeled substrates accumulated in an intracellular glutamate pool. Glutamate concentrations increase when the external osmotic pressure is raised (Hua et al., 1982) and the existence of such a pool may be an important complication in analysis of Rhizobium carbon metabolism.

We have attempted to determine how glutamate is catabolized by following the fate of radioactive glutamate in whole cells and in cell extracts and by isolating Tn5 mutants of R. meliloti 104A14 that grow poorly on glutamate as sole carbon and nitrogen source but that grow normally on mannitol/ammonia or arabinose/ammonia minimal medium. Our results indicate that an NAD dependent glutamate dehydrogenase activity is important in glutamate degradation. There are several lines of evidence that support this conclusion:

1. There is a substantial glutamate and NAD dependent dehydrogenase (GDH-NAD) activity in our strain (Fig 1). In freshly prepared extracts using high concentrations of substrates, this activity is masked by an equally active NADH dehydrogenase activity. The latter can be inhibited by the addition of the respiratory chain inhibitors rotenone or antimycin A (De Hollander and Stouthamer, 1980).

2. The consumption of radioactively labeled glutamate by whole cells was inhibited by 1 mM sodium arsenite, an inhibitor of the lipoic acid containing enzyme a-ketoglutatrate dehydrogenase. CO_2 release is blocked completely and analysis of the soluble products of this degradation by thin layer anion exchange chromatography shows that inside the cells there is essentially no conversion of the glutamate label to other compounds but that a-ketoglutarate accumulates in the medium. In cell free extracts with low concentrations of substrates, the predominant pathway of glutamate degradation is through gamma-aminobutyric acid but this pathway is not blocked by sodium arsenite at moderate concentrations.

3. One Tn5 mutant, R. meliloti 104A14 gca90, grew very slowly on glutamate but grew well on the other media above. This mutant lacked normal GDH-NAD activity. The mutant was also inhibited in its growth on gamma amino butyric acid, aspartate and proline and, unlike wild type, did not give an alkaline reaction when grown on a mannitol ammonia medium that contains high concentrations of either glutamate or aspartate. Revertants that grow well on glutamate are readily isolated but both the original Tn5

mutation and the GDH defect are usually still present and these isolates are therefore probably pseudorevertants. Plants infected with the mutant are Fix+ but revertants (or pseudorevertants) can always be found associated with these plants. The revertants produced a higher than normal frequency of Fix- plants. The wild type gene that complements mutant gca90 has been cloned but we have not yet been able to determine whether it carries a structural gene for NAD-GDH or a regulatory gene.

This work was supported by grants from the United States Department of Agriculture and the Washington Technology Center.

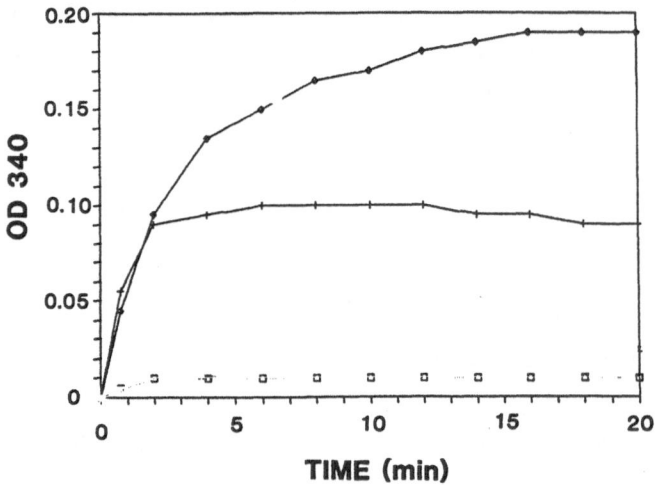

TIME (min)

Fig. 1. Glutamate dependent dehydrogenase activity in R. meliloti 104A14. Cells were grown in minimal glutamate medium, concentrated 100 fold and lysed by sonication. The lysate was centrifuged at 15,000 g for 10 min. 20 ul of the supernatant was added to 1.5 ml of 100 mM Tris, 83 mM sodium glutamate, 0.85 mg/ml NAD pH 7.5. The increase in absorbance at 340 nm at 22°C is the result of the production of NADH.
(□) no addition; (◊) + 12 uM antimycin A; (+) + 20 uM rotenone.

Carlson, T.A., Martin, G.B., and Chelm, B.K. 1987. Differential transcription of the two forms of glutamine synthetase genes of Bradyrhizobium japonicum. J. Bacteriol. submitted.

De Hollander, A. and A.H. Stouthamer. 1980. The electron transport chain of Rhizobium trifolii. Eur. J. Biochem. 111, 473-478.

Dilworth, M.J. and Glenn, A.R. 1985. In "Nitrogen Fixation and CO_2 Metabolism" (Ed. P.W. Ludden and J.E. Burris). (Elsevier, New York). pp. 53-61.

Donald, R.G.K. and Ludwig, R.A. 1984. Rhizobium sp. Strain ORS571 ammonium assimilation and nitrogen fixation. J. Bacteriol. 158, 1144-1151.

Emerich, D.W., Lepo, J.E., and Evans, H.J. 1983. Nodule metabolism. In "Nitrogen Fixation, Vol 3: Legumes" (Ed. W.J. Broughton). (Clarendon Press, Oxford). pp. 213-244.

Hua, S.-S.T., Tsai, V.Y., Lichens, G.M. and Noma, A.T. 1982. Accumulation of amino acids in Rhizobium sp. strain WR1001 in response to sodium chloride salinity. Appl. Environ. Microbiol. $\underline{44}$, 135-140.

Kahn, M.L., Kraus, J., and Somerville, J.E. 1985. A model for nutrient exchange in the Rhizobium-legume symbiosis. In "Nitrogen Fixation Research Progress" (Ed. H.J. Evans, P.J. Bottomley, and W.J. Newton). (Martinuus Nijhoff, Dordrecht). pp. 193-199.

Kohl, D.H., Schubert, K.R., Carter, M.B., Hagedorn, C.H. and Schearer, G. 1987. Proline metabolism in N_2-fixing root nodules: energy transfer and regulation of purine synthesis. submitted.

Kumar, P.S. and Rao, S.L.N. 1986. Identification and characterization of three forms of glutamine synthetase unique to rhizobia. Curr. Microbiol. $\underline{14}$, 113-116.

Murphy, P.J., Heycke, N., Banfalvi, Z., Tate, M.E., de Bruijn, F., Kondorosi, A., Tempe, J. and Schell, J. 1987. Genes for the catabolism and synthesis of an opine-like compound in Rhizobium meliloti are closely linked and on the Sym plasmid. Proc. Nat. Acad. Sci. (U.S.A.) $\underline{84}$, 493-497.

O°Gara, F., Manian, S. and Meade, J. 1984. Isolation of an Asm⁻ mutant of Rhizobium japonicum defective in symbiotic N_2 fixation. FEMS Microbiol. Lett. $\underline{24}$, 241-245.

Richey, B., Cayley, D.S., Mossing, M.C., Kolka, C., Anderson, C.F., Farrar, T.C. and Record, M.T., Jr. 1987. Variability of the intracellular ionic environment of Escherichia coli. J. Biol. Chem. $\underline{262}$, 7157-7164.

Ronson, C.W. and Astwood, P.M. 1985. Genes involved in the carbon metabolism of bacteroids. In "Nitrogen Fixation Research Progress" (Ed. H.J. Evans, P.J. Bottomley, and W.J. Newton). (Martinuus Nijhoff, Dordrecht). pp. 193-199.

Salminen, S. O. and Streeter, J. G. 1987. Involvement of glutamate in the respiratory metabolism of Bradyrhizobium japonicum bacteroids. J. Bacteriol. $\underline{169}$, 495-499.

STUDIES ON THE RHIZOBIUM-PEA SYMBIOSIS: THE ROLE OF MALATE
DEHYDROGENASE AND THE EFFECT OF THE PLANT ON THE
EFFICIENCY OF THE NITROGENASE REACTION

H. Haaker, M.A. Appels and J.H. Wassink

Department of Biochemistry
Agricultural University
De Dreijen 11, 6703 BC Wageningen
The Netherlands

ABSTRACT

    Two aspects of the metabolism in pea root nodules in relation to
nitrogen fixation have been studied. The role of malate dehydrogenase
in the carbon metabolism and the efficiency of the nitrogenase reaction in intact root nodules was investigated. The presence of a root
nodule stimulated form of malate dehydrogenase is demonstrated. From a
comparison of the kinetic properties of this enzyme and the enzyme
from root cells, it is concluded that the root nodule form of malate
dehydrogenase is capable of allowing a high flux of malate production
from oxaloacetate and also to establish a sufficient oxaloacetate concentration necessary for the synthesis of amino acids involved in the
transport of newly fixed nitrogen from the root nodules. Data are presented to demonstrate that the variable H_2 production by nitrogenase
in intact root nodules is not caused by O_2 limitation of the bacteroids but by plant factors like free fatty acids.

INTRODUCTION

    Most of the Rhizobia and Bradyrhizobia species that infect legumes

and form an effective symbiosis do not fix dinitrogen ex planta. Only

under micro-aerophilic conditions can the nif genes be derepressed and

nitrogen fixation activity demonstrated. The whole cell nitrogenase

activity of Rhizobia is in most cases not high enough to allow growth

with N_2 as N-source. Diazotropic growth has been reported only for a

limited number of Rhizobia, (Bergersen et al., 1976, Dreyfus and

Dommerques, 1981, Stam et al., 1984). It is generally accepted that in

the symbiosis the plant generates, in the root nodules, an environment

suitable for nitrogen fixation by the micro-symbiont. In case of the

Rhizobium-legume symbiosis, this is a low free oxygen concentration

(<1 μM) and an addequate supply of C4-dicarboxylic acids (Streeter and

Salminen, 1985). The Rhizobium bacteroids cannot assimilate the newly

F. O'Gara et al. (eds.), Physiological Limitations and the Genetic Improvement of Symbiotic Nitrogen Fixation, 29–40.
© 1988 by Kluwer Academic Publishers.

fixed nitrogen and NH_3 diffuses to the plant cytoplasm where it is assimilated in a reaction catalized by glutamine synthase. During the study of the legume symbiosis it became clear that symbiotic nitrogen fixation is a highly developed process and involves a differentiation of the plant and the bacterium. Despite this differentiation the symbiosis is not optimal and the efficiency can be improved. In this paper two aspects of symbiotic nitrogen fixation will be discussed. The first topic deals with the generation of oxidizable substrates for the bacteroids and carbon skeletons for the synthesis of amino acids used for export of fixed nitrogen to other parts of the plant. In the second part of this paper the question whether the conditions in the cytoplasm of root nodule cells are always optimal for nitrogen fixation by the bacteroids will be discussed .

MATERIALS AND METHODS

Growth conditions of the plants and preparation of extracts

Plants were grown and bacteroids were isolated as described earlier (Haaker and Wassink, 1984). Nodules were cut from the main root at 18-19 days after inoculation. Uninfected plants were cultured in the same way as infected plants but with sufficient nitrogen, supplied as NH_4NO_3 (0.025 g per liter). The main roots of uninfected plants were harvested and the parts containing meristematic tissue were removed. From roots with developing nodules, a 2.5 cm piece of the main roots was cut at the location where normally the nodules appear. To prepare extracts, root nodules or roots were ground in a chilled mortar with isolation buffer (approx. 1 ml/g fresh material). The isolation buffer consisted of 50 ml TES-KOH, 16% w/v sucrose, 0.9% w/v glucose, 5 mM DTT and 1 mM EDTA, pH 7.4 at 4°C. After filtering the homogenate through Miracloth (Calbiochem), bacteroids were removed by centrifugation (20 min at 6000 x g). The resulting supernatant contained soluble cytoplasmic plant proteins and cell organelles, mainly mitochondria and microsomes. The soluble cytoplasmic plant proteins were separated from cell organelles by high speed centrifugation (80 min at 60,000 x g).

Analytical methods

Desalted samples were analyzed using a Mono Q column (bed volume
1 ml; FPLC-system of Pharmacia) equilibrated with 20 mM Tris-HCl, pH
7.4. About 1% of the protein load capacity of the column was utilized,
based on the information given by Pharmacia. The proteins were eluted
at 0.5 ml/min with a NaCl-gradient. The increase in the NaCl concentration was 10 mM/ml. Fractions of 0.5 ml were collected and analyzed for malate dehydrogenase activity.

Nitrogenase activity (C_2H_2 reduction) and the electron allocation
coefficient were determined as described earlier (Haaker and Wassink,
1984). Malate dehydrogenase activity (E.C. 1.1.1.37) (oxaloacetate
reduction) was measured at 25 °C in 25 mM potassium phosphate buffer,
0.2 mM NADH and 0.4 mM oxaloacetaat, pH 7.4. Malate oxidation catalyzed by malate dehydrogenase was measured in the presence of glutamate oxaloacetate transaminase (E.C. 2.6.1.1.) (4.0 U/ml), 50 mM
Tris-HCl, 100 mM L-malate, 0.8 mM NAD^+, 40 mM L-glutamate, pH 8.0.

RESULTS AND DISCUSSION

THE ROLE OF MALATE DEHYDROGENASE IN THE PEA NODULE

The predominant C4-dicarboxylic acid in root nodules of legumes is
malate (Stumpf and Burris, 1979), which is supposed to be synthesized
from NADH and oxaloacetate. This reaction is catalyzed by malate
dehydrogenase in the cytoplasm of the root nodule cells. The activity
of this enzyme is high in root nodules (De Vries et al., 1980, Henson
and Collins, 1984). Under standard conditions the reduction of oxaloacetate by NADH is exergonic ($\Delta G^{o'}=-2.73 \times 10^4$ J). Thus under physiological conditions, at pH=7.0-7.3 and $NADH/NAD^+=0.2-0.4$, the
equilibrium of the reaction catalyzed by malate dehydrogenase is
completely in favour of NAD^+ and malate. In equilibrium at a ratio of
$NADH/NAD^+=0.3$ and 4 mM malate the oxaloacetate concentration is 0.2
μM. Thus with an active malate dehydrogenase present, a substantial
reduction of oxaloacetate to malate is to be expected. Oxaloacetate is
not only used in the reaction catalyzed by malate dehydrogenase but also
acts as a carbon skeleton used for the assimilation and transport of

fixed nitrogen to other parts of the plant (Boland et al., 1980).
Oxaloacetate is used in the reaction catalyzed by glutamate oxaloacetate transaminase (E.C. 2.6.1.1.) to produce aspartate. The K_m value
for oxaloacetate has been estimated to be 110 μM for soybean (Ryan et
al., 1972) and 20 or 100 μM for lupine depending on the type of
isoenzyme (Reynolds et al., 1981). These values indicate that a concentration of oxaloacetate above 10 μM is a requirement for a reasonable rate of assimilation of fixed nitrogen via the oxaloacetate,
aspartate and asparagine pathway (Scott et al. 1976). Despite the high
malate dehydrogenase activity in root nodules asparagine synthesis
takes place. To solve this problem and to give an answer to the
question of the role of malate dehydrogenase in symbiosis, malate
dehydrogenase from the cytoplasm of root nodules cells was studied and
compared with the enzyme present in the cytoplasm of root cells.

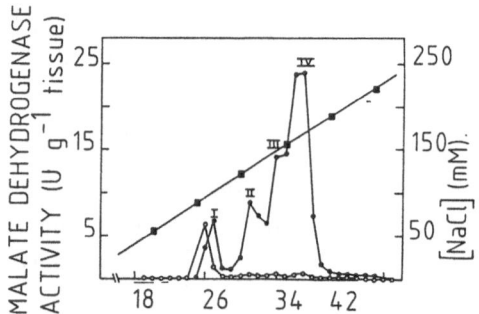

Figure 1. Elution profile of malate dehydrogenase activity from Mono Q
of the cytoplasmic plant fraction of pea root nodules harvested 19
days after inoculation (•——•) and of the cytoplasmic fraction of the
uninoculated roots harvested 17 days after sowing (○——○). The malate
dehydrogenase activity was measured by the reduction of oxaloacetate.
NaCl concentrations were determined in the fractions (■——■).

In fig.1 the elution patterns of an anion exchange column are shown.
The fractions of the column were analyzed for the presence of malate
dehydrogenase. Cytoplasmic fractions of root nodule cells and root
cells were applied to the ion-exchange column. The elution profile
of malate dehydrogenase of the cytosolic fraction of root cells show
one main peak at 95 mM NaCl and three peaks of minute activity at
higher NaCl concentrations. The elution pattern of malate dehydrogenase present in the cytoplasm of root nodule cells is different. The
first peak (I) elutes at a slightly higher NaCl concentration as the

main root fraction. Furthermore the main activity elutes in three other peaks at higher NaCl concentrations (130–180 mM NaCl, II, III and IV). These fractions account for more than 90% of the total activity eluted from the column. The behavior of malate dehydrogenase on the ion-exchange column was also studied during nodule development. As

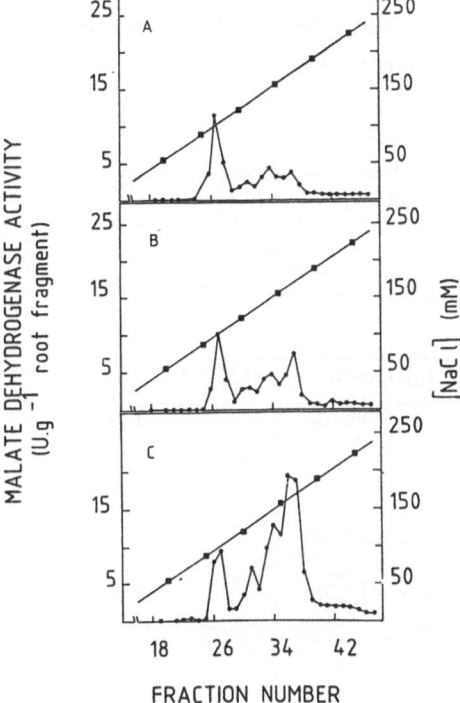

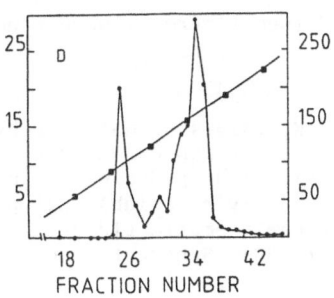

Figure 2. Elution profile of malate dehydrogenase activity from Mono Q of root fragments with developing nodules harvested at different times following inoculation of Pisum sativum with Rhizobium leguminosarum PRE. ●——●, the malate dehydrogenase activity: a, cytoplasmic fraction of root fragments harvested 8 days after inoculation. b, 15 days; c, 19 days; d, 25 days. The NaCl concentration of the fractions (■——■).

can be seen in figure 2, 8 days after inoculation root cell differentiation to root nodule cells can be detected biochemically by ion-exchange chromatography. At this time after inoculation no nodules are visable on the roots, therefore root fragments where nodules normally appear were used. At day 11 nitrogenase activity could be detected. These results suggest that the root nodule stimulated forms of malate dehydrogenase (fractions II, III and IV of Fig. 1) appear at the same time as the early nodulines in pea are detectable (Govers et al. 1985). Experiments were carried out to check that the root nodule forms of malate dehydrogenase are not from bacterial, bacteroid, meristematic or mitochondrial origin. This was done by measuring the elution patterns of malate dehydrogenase of the soluble fractions of the different preparations mentioned (not shown). We also measured

the isoelectric points of the separated malate dehydrogenases. In combination with the elution patterns it could be concluded that the root
nodule fractions II, III and IV are not from bacterial, mitochondrial,
meristematic or bacteroid origin. The pI, Mr, and the elution pattern
from the ion-exchange column indicated that root nodule fraction I
might originate from meristematic tissue.

Some kinetic properties of the main root and root nodule fractions
I and IV (see figure 1) were studied. The results are presented in
table 1.

Table 1 K_m-values of malate dehydrogenase fractions. Fractions were
collected after ion-exchange chromatography on Mono Q of the soluble
cytoplasmic plant proteins of nodules (root nodule fraction I and IV)
and after applying soluble cytoplasmic proteins of uninoculated main
roots or meristematic tissue. Data were analyzed as described by
Dalziel (1969). Substrate coefficients were estimated at 99% level of
significance.

Fraction	K_m(malate)	K_m(NAD$^+$)	K_m(oxaloacetate)	K_m(NADH)
root nodule fraction I	9.7 mM	1.4 mM	65 μM	39 μM
root nodule fraction IV	2.6 mM	27 μM	18 μM	13 μM
main root fraction	64 mM	4.4 mM	89 μM	70 μM
meristematic	7.8 mM	0.93 mM	94 μM	63 μM

The K_m-values for all four substrates of malate dehydrogenase present
in root nodule fraction IV were lower than those of the main root
fraction. The K_m-values of root nodule fraction I were in between
those of the two mentioned preparations. The possibility that regulation of the reduction of oxaloacetate occurs by the formation of an
abortive complex of malate dehydrogenase with malate and NADH which is
inactive was examined. Therefore the kinetic data of the different
malate dehydrogenases were analyzed by a 2-substrate compulsory order
ternary-complex mechanism. This is the generally accepted mechanism
for soluble malate dehydrogenases. The cofactor NAD$^+$ or NADH must bind
first to the enzyme. After binding of the cofactor the other
substrate, oxaloacetate or malate, will bind and catalysis can take
place. When malate binds to the malate dehydrogenase-NADH complex a
so-called abortive complex is formed, which is inactive. As discussed
by Dalziel (1969), Φ-parameters are indicative for the formation of

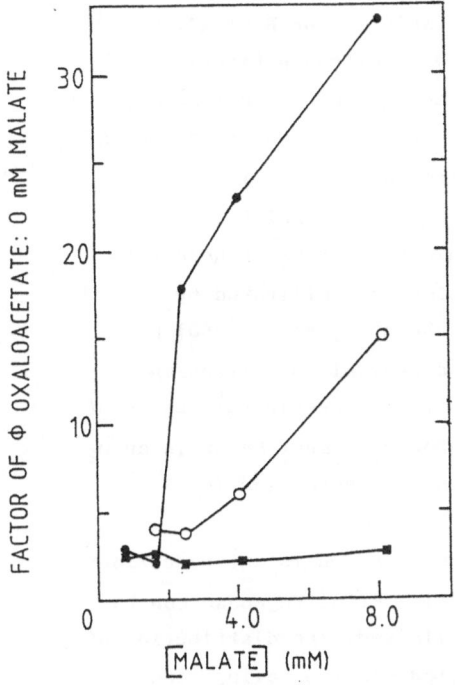

Figure 3. Formation of the abortive complex malate dehydrogenase-NADH-malate measured by the Dalziel substrate coefficient $\Phi_{oxaloacetate}$ for root nodule fraction I (○——○), root nodule fraction IV (●——●) and for the main root fractions (■——■). The substrate coefficient $\Phi_{oxaloacetate}$, which is a parameter for the abortive complex mentioned, was determined as described by Dalziel (1969) in the presence of different malate concentrations. The substrate coefficient is expressed as factor of the $\Phi_{oxaloacetate}$ in the absence of malate.

abortive complexes. As can be seen in fig. 3, the $\Phi_{oxaloacetate}$ of malate dehydrogenase fraction I and IV of root nodules increases with the malate concentration, while that of the main root fraction is constant. The other Φ-parameters of all fractions remain constant or increase within the limit of a factor three (not shown). These results show that at physiological concentrations malate can act as a regulator of the activity of the root nodule forms of malate dehydrogenase.

In conclusion, a role of the cytoplasmic malate dehydrogenase in symbiosis can be suggested. A high activity of malate dehydrogenase for the reduction of oxaloacetate to malate is important for the high demand for malate as oxidizable substrate for the bacteroids. When the malate concentration in the cytoplasm is in the mM range (3-7 mM), the activity of the enzyme is inhibited due to the formation of an abortive complex. This prevents the enzyme from catalyzing the reaction to equilibrium, which would otherwise lead to an oxaloacetate concentration too low for the synthesis of amino acids involved in the transport of fixed nitrogen to other parts of the plant.

ELECTRON ALLOCATION TO H^+ AND N_2 BY NITROGENASE IN THE BACTEROIDS

Oxygen plays a dualistic role in symbiotic nitrogen fixation.
Oxygen is necessary for the generation of energy in the bacteroids,
since bacteroids are obligate aerobes, but at higher concentrations of
O_2 low potential electron carriers and nitrogenase are oxidized.
Several mechanism are operative in root nodules to regulate the O_2
concentration around the bacteroids. A rapid rate of O_2 transport to
the bacteroids at a low free O_2 concentration is facilitated by
leghaemoglobin (Wittenberg et al., 1974). Sheehy et al., (1983) proposed another regulating mechanism namely a variable O_2 diffusion
resistence. These two mechanism together with the respiratory rates of
O_2 consumption by the bacteroids and mitochondria, must maintain an O_2
concentration at which nitrogen fixation in the bacteroids is
possible.

Another factor that influences nitrogen fixation is the reduction
of H^+ to H_2 by nitrogenase. This side reaction of nitrogenase can be
very important when the plant growth is N-limited. The distribution of
electrons at the nitrogenase enzyme is called electron allocation or
the relative efficiency of the nitrogenase. The electron allocation
coefficient is defind as electrons used to reduce N_2/total electron
flux through nitrogenase. The conditions determining the allocation of
electrons to N_2 and H^+ have been extensively studied with the isolated
enzyme (Hageman and Burris, 1980). All variables that influence the
flux of electrons through nitrogenase component I have the same effect
on the electron allocation. When the flux of electrons was increased,
N_2 was favoured as a substate by nitrogenase above H^+. When nitrogenase was inhibited H^+ was favoured as substrate above N_2. Haaker and
Wassink (1984) studied electron allocation in intact bacteroids. They
showed that there are important differences between the behaviour of
the enzyme in vitro and in vivo. In vivo the electron allocation was
only influenced by the intracellular ADP concentration. When the ADP
concentration is relatively high, nitrogenase is partly inhibited and
a relative large part of the electrons taken up by nitrogenase are
used for H^+ reduction. This is the case when isolated bacteroids are

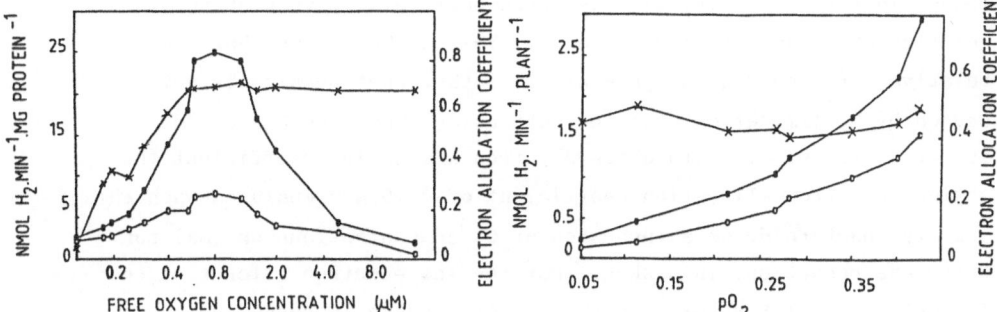

Figure 4. The effect of oxygen on the relative efficiency of the nitrogenase reaction in isolated bacteroids and intact root nodules of pea. The experiments have been performed as described earlier (Haaker and Wassink, 1984). A, isolated bacteroids; B, detached pea roots, per datum point at least 20 plants were used. •——•, H_2 production under Ar/O_2; o——o, H_2 production under N_2/O_2; X——X, electron allocation coefficient.

incubated under O_2-limited conditions (see fig. 4a). Under these conditions the electron allocation coefficient is low (0.05-0.4)). The same phenomenon was observed when the cells were incubated with proton conducting uncouplers of oxidative phosphorylation. Electron allocation to H^+ was not stimulated when electron transport to nitrogenase was inhibited. Due to inhibition by excess O_2 (see fig. 4a). When the nitrogenase activity of the bacteroids is stimulated by a higher O_2 input rate, the nitrogenase activity and the electron allocation coefficient increase to their maximal values.

When the electron allocation coefficient of nitrogenase was measured in intact pea plants relatively low values were found : 0.4 ±0.16. From the results of the isolated bacteroids one would expect that when the O_2 supply to the bacteroids is increased one would observe a higher rate of nitrogense activity and a higher electron allocation coefficient. This was not observed (see figure 4b). When the pO$_2$ was increased the H_2 production under N_2/O_2 mixtures and the H_2 production under Ar/O_2 mixtures were both stimulated to the same

extent. Thus while the nitrogenase activity increases by a factor of
10, the relative efficiency of nitrogenase hardly changed. A similar
change in total activity with isolated bacteroids gave a change in
relative efficiency from 0.04 to 0.7. This demonstrates that the low
relative efficiency of nitrogenase in intact root nodules is not
caused by O_2 limitation near the bacteroids. When the bacteroids are
isolated from plants with a low electron allocation coefficient the
maximal electron allocation coefficient of 0.75 was measured when the
isolated bacteroids were incubated at O_2 concentrations optimal for
acetylene reduction. This shows that the low electron allocation coefficient (low relative efficiency) of nitrogenase is caused by the
environment of the bacteroids in the cytoplasm of the plant. Laane et
al. (1978) observed that the nitrogenase activity of freshly isolated
Rhizobium leguminosarum bacteroids could be considerably stimulated by
the addition of fatty acid free bovine serum albumin. It was shown
that oxidative phosphorylation and nitrogen fixation of isolated bacteroids could be inhibited by low amounts of free fatty acids (10-25
µg linoleic acid/mg protein). It might be possible that the presence
of compounds like free fatty acids in the cytoplasm of root nodule
cells influences nitrogen fixation by lowering the relative efficiency
of the nitrogenase reaction. This would explain the experiments where
the relative efficiency of the nitrogenase reaction is low and independent of the absolute activity. When the O_2 supply to such nodules
is increased more bacteroids start to fix N_2 with the same low relative efficiency. When the oxygen supply is limiting nitrogen fixation
of the bacteroids one expects an increase in relative efficiency
together with in increase in absolute nitrogenase activity.

ACKNOWLEDGEMENTS

The present investigation was supported by the Netherlands
Foundation for Chemical Research (SON) with financial aid from the
Netherlands Organization for the Advancement for Pure Research (ZWO)
and by a grant from the EC under contract no. 1915-NL.

REFERENCES

Bergersen, F.J., Turner, G.L., Gibson, A.H. and Dudman, W.F. 1976.
Nitrogenase activity and respiration of cultures of Rhizobium spp.
with special reference to the concentration of dissolved oxygen.
Biochem. Biophys. Acta 444, 164-174.

Boland, M.J., Farnden, K.J.F. and Robertson, J.G. 1980. Ammonia
assimilation in nitrogen fixing legume nodules. In "Nitrogen fixation
Vol. 2" (Eds. W.E. Newton and W.H. Orme-Johnson) University Park Press
Baltimore, pp 33-52.

Dalziel, K. 1969. The interpretation of kinetic data for
enzyme-catalyzed reactions involving three substrates. Biochem.J. 114,
547-556.

DeVries, G.E., In 't Veld, P. and Kijne, J.V. 1980. Production of
organic acids in Pisum sativum as a result of oxygen stress. Plant
Sci. Lett. 20, 115-123.

Dreyfus, B.L. and Dommergues, Y.R. 1981. Nitrogen fixing nodules
induced by Rhizobium on the stem of the tropical legume Sesbania
rostrata. FEMS Microbiol Lett., 5, 369-372.

Govers, F., Gloudemans, T., Moerman, M., Van Kammen, A. and Bisseling,
T. 1985. Expression of plant genes during nodule development of pea
nodules. EMBO J., 4, 861-867.

Haaker, H. and Wassink, H. 1984. Electron allocation to H+ and N2 by
nitrogenase in Rhizobium leguminosarum bacteroids. Eur. J. Biochem.
142, 37-42.

Hageman, R.V. and Buris, R.H. 1980. Electron allocation to alternative
substrates of Azotobacter nitrogenase is controlled by the electron
flux through dinitrogenase. Biochim. Biophis. Acta 591, 63-75.

Henson, C.A. and Collins, M. 1984. Carbon metabolism in alfaalfa root
nodules: developmental patterns of host plant enzymes before and after
shoot removal. Crop. Sci. 24, 727-732.

Laane, C., Haaker, H. and Veeger, C. 1978. Involvement of the
cytoplasmic membrane in nitrogen fixation by Rhizobium leguminosarum
bacteroids. Eur.J.Biochem. 87, 147-153.

Reynolds, P.H.S., Boland, M.J. and Farnden, K.J.F. 1981. Enzymes of
nitrogen metabolism in legume nodules : partial purification and
properties of the aspartate aminotransaminase from lupine nodules.
Arch. Biochem. Biophys. 209, 524-533.

Ryan, E.D., Bodley, F. and Fottrell, R.F. 1972. Purification and
characterization of aspartate aminotransaminase from soybean root nodules
and Rhizobium japonicum. Phytochemistry 11, 957-963.

Scott, D.B., Farnden, K.J.F. and Robertson, J.G. 1976. Ammonium
assimilation in lupin nodules. Nature 263, 703-705.

Sheeny, J.E., Minchin, F.R. and Witty, J.F. 1983. Biological control of the resistance to oxygen flux in nodules. Annals of Botany 52, 565-571.

Stam, H., van Verseveld, H.W., de Vries, W. and Stouthamer, A.H. 1984. Hydrogen oxidation and the effiency of nitrogen fixation in succinate-limited chemostat cultures of Rhizobium OR 571. Arch. Microbiol., 139, 53-60.

Stumf, D.K. and Burris, R.H. 1979. A micromethod for purification and quantification of organic acids of the tricarboxylic acid cycle in plant tissue. Anal. Biochem. 95, 311-315.

Streeter, J.G. and Salminen, S.O. 1985. Carbon metabolism in legume nodules. In "Nitrogen fixation research progress"(Eds.H.J. Evans, P.J. Bottomly and W.E. Newton). Martinus Nijhoff Publishers Dordrecht. pp. 277-283.

Wittenberg, J.B., Bergersen, F.F., Appleby, C.A. and Turner, G.L. 1974. Facilitated diffusion of O2 : the role of oxyleghaemoglobin in nitrogen fixation by bacteroids isolated from soybean root nodules. J. Biol. Chem. 249, 4057-4066.

ENERGETICS OF SYMBIOTIC NITROGEN FIXATION :

THE RELATIONSHIP BETWEEN OXYGEN, MALATE AND HYDROGEN

J.J.DREVON, A.GODFROY, M.O. HECKMANN, V.C.KALIA, N.OLLAT
Institut National de la Recherche Agronomique
Laboratoire de Recherche sur les Symbiotes des Racines
Montpellier, 34000, France

ABSTRACT
    The energetics of symbiotic nitrogen fixation was investigated by measuring on intact soybean plant, with an open flow device, the influence of the root medium oxygen and malate content on the nodule nitrogenase acetylen reducing activity. It was shown that the maximum nodule activity required more oxygen than in air and was higher in presence of malate, but the photosynthesis modification by light intensity manipulation had no effect on nodule activity, although it affected the total mass of nodule per plant. The presence of an hydrogenase, which oxidized the hydrogen produced by the nitrogenase, had no beneficial effect on nitrogen fixation and growth of soybean, probably because of the oxygen limitation of nodule activity. In the absence of hydrogenase, the hydrogen production was higher than the minimum one measured on the purified enzyme ; it varied with the environment and the host-plant cultivar.

INTRODUCTION

    The reduction of atmosphere nitrogen into ammonium by the nitrogenase enzyme requires a lot of energy in reducing power and ATP (Salsac et al., 1984). This high energetic cost of nitrogen fixation is increased by the nitrogenase synthesis of hydrogen, a gas which is generally evolved by root legume nodules except with the few strains of Rhizobium that possess an active hydrogenase (Eisbrener and Evans, 1983). Other major expenditures associated with the symbiotic N_2 fixation are the requirements for the nodule development and maintainance, and the carbon skeletons, reducing power and ATP necessary for the ammonium assimilation. Consequently, the root-nodule operation requires important fluxes of carbon substrates for reducing power and oxygen for ATP generation. These fluxes are mediated respectively by the host-plant supply of sucrose from its photosynthesis and a facilitated O_2 diffusion inside the nodule (Sheehy et al., 1985).

    This paper presents the results of *in situ* assays of nodule acetylen reduction in presence of various oxygen partial pressures. The influence of the carbon flux was studied by comparing plants grown under various illuminations or by supplying carbon substrates in the root medium.   The energetic influence of the hydrogenase and the magnitude of the hydrogen production by the nitrogenase is also addressed.

F. O'Gara et al. (eds.), Physiological Limitations and the Genetic Improvement of Symbiotic Nitrogen Fixation, 41–50.
© 1988 by Kluwer Academic Publishers.

OPTIMAL EXTERNAL OXYGEN TENSION

To study the influence of the external partial pressure of O_2 on the symbiotic nitrogenase activity, nodulated soybeans were grown hydroponically in a controlled environment and assayed in the open-flow device described in figure 1, in which various compositions of the gas flow could be monitored.

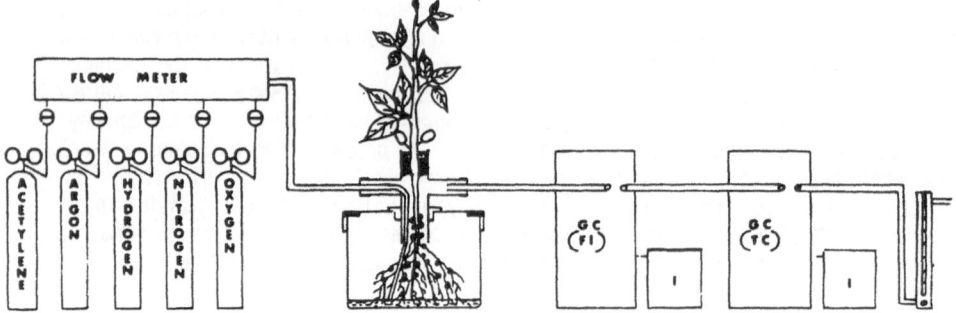

Fig. 1 Device for the open-flow in situ assay of legume nodule activity (Drevon et al., 1988 a).

It was shown initially that on these plants grown in liquid aerated medium and incubated under 700 µmol photons m^{-2} s^{-1} and a steady temperature, the conventional C_2H_2 reductionn assay (Hardy et al., 1973) was not inducing nitrogenase inhibition (Drevon et al., 1988 a) in contrast with the phenomenon described on nodulated legumes grown on perlite (Minchin et al., 1983).

Since the ARA (acetylen reducting activity) maintained steady during at least 8 hours if the environment was not modified, it was possible to monitor on a single plant the response of the ARA to raising O_2 concentrations in the gas-flow (Drevon et al., 1987 a). The result of such an experiment presented in figure 2 revealed an optimal external O_2 tension for the nodule nitrogenase activity which was close to 30% O_2 for young nodules of soybean C.V.Kinsoy, and closer to 20% O_2 for older nodules. It also confirmed the existence of some control of the O_2 diffusion at supraoptimal O_2 tensions, since the ARA declined only slightly or was maintained at these O_2 concentrations ; otherwise the nitrogenase would have been rapidly inactivated by an excess of oxygen in its environment.

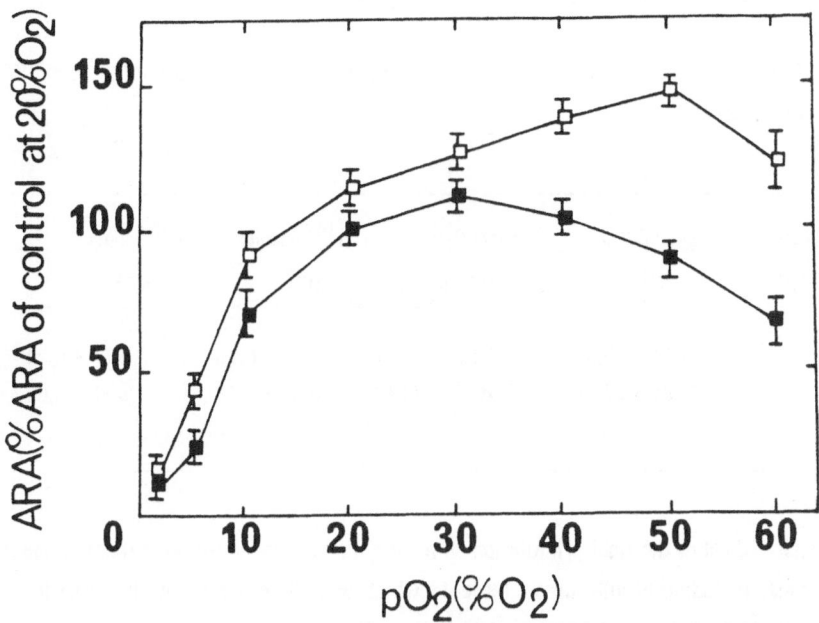

Fig.2 Influence of the external O_2 partial pressure on the nodule nitrogenase acti(C_2H_2 reduction)in absence (- -) or in presence (- -)of 15 mM in the root medium.

The profile of the ARA response to O_2 was also shown to vary with nitrate supply (Heckmann et al., this volume) and recently with the host-plant cultivar (Seraj et al., unpublished data). Therefore, these data which agree with previous reports of short term experiments on intact palnts (Criswell et al., 1977 ; Ralston and Imsande, 1982) suggest that in natural habitats, the root-legume nodule nitrogenase activity would be limited at least in some symbiosis and at some stages of the growth cycle by the O_2 pressure in the root atmosphere which could not be more than 21%. However it has to be known if the nodule could maintain their activity at optimal supraambient O_2 tension for more than a few hours.

PHOTOSYNTHESIS AND MALATE SUPPLY

Since the O_2 supply limits nitrogenase activity, the host plant supply of photosynthate might not be a limiting factor of the nodule activity.Thus, nodules from soybeans grown under various illuminations from 200 to 700 µmol photons $m^{-2} s^{-1}$ had similar specific activity, i.e. ARA per gram fresh weight nodules, although total nitrogenase activity was much higher at the optimal illumination of 700 µmol photons $m^{-2} s^{-1}$ (Table 1).

Table 1. Total and specific nitrogenase activity of plants grown under optimal or suboptimal illuminations.

µmol photons $m^{-2} s^{-1}$	µmol C_2H_4 $h^{-1} g^{-1}$ FN nod	mmol C_2H_4 $h^{-1} pl^{-1}$	g ^{DW} nodules pl^{-1}	g ^{DW} shoots pl^{-1}
400	89,71 + 19,32	21,39 + 6,35	0,25 + 0,03	3,56 + 0,21
700	82,62 + 26,05	48,22 + 11,80	0,59 + 0,06	6,32 + 0,65

The major effect of the light stimulation on photosynthesis was to increase the total mass of plant tissue, including nodule mass (Table 1) which was the cause of the increase in total nitrogen fixation. But it had no apparent effect on the nodule activity per se. Moreover, a limitation of the light intensity on the plants grown at the optimal illumination had no detectable effect on the nodule ARA at 21% O_2 during the next 48 h (Drevon et al., 1988 b). An increase of photosynthesis by raising the CO_2 partial pressure in the shoot atmosphere neither had any short term effect on the nodule ARA (Finn and Brun, 1982).

Therefore, it appears that in air the nodule activity is not limited by the concurrent photosynthesis and that it can rely on photosynthate reserves at least for 2 days in soybean. The symbiosis responds to an increase in shoot photosynthesis and N demand by the development of new nodules rather than by the activation of the existing nodules.

The absence of short term nodule response to the photosynthesis increase in the above assay performed in the presence of 21% O_2 might have been caused by the O_2 limitation and the incapacity of the nodule to increase their O_2 diffusion capacity. In contrast, at supra optimal O2 tension, the nodule activity was highly stimulated by an external supply of carbon substrate under the form of malate (Fig.2) : the optimal ARA was 150% higher than the control one ; the optimal oxygen tension was increased to 50% O_2 . Even at 20% O_2, the malate supply stimulated the ARA, although an equivalent supply of glucose or sucrose had no effect (Table 2).

Table 2. Effect of external carbon supplu on soybean nodule nitrogenase activity at 20% O_2

---

control	malate	glucose	sucrose	CO_2
$100,0 \pm 4.2$	$133,4 \pm 6.7$	$99,5 \pm 2.1$	$98,0 \pm 5.1$	$154,2 \pm 6.9$

Thus, the bacteroids could use various substrates in the soybean symbiosis and preferentially malate which would be among the most efficient source of reducing power ; but the rate of malate synthesis in nodule would not be sufficient to meet the bacteroids requirements in reducing power. A possible major pathway of malate synthesis is the reduction of the oxaloacetate produced by the phospho-enol pyruvate (PEP) carboxylation (Fig.3). Indeed the two corresponding enzymes, i.e. malate dehydrogenase and PEP carboxylase are present in large amount in the nodule cytosol (Coker and Schubert,1981 ; Vance et al., 1983) and the ARA responds to the external CO_2 pressure (Bethenod et al., 1984 ; Table 2).

Whatever the specificity of malate, its effect on the profile of the ARA response to O_2 indicates that the level of optimal oxygen tension depends upon the reducing power availability in the nodules. Thus the profile would be an indicator not only of the O_2 limitation but also of the carbon supply of the nodule bacteroids.

INFLUENCE OF THE HYDROGENASE

Some *Rhizobium* strains having a hydrogenase, consequently called Hup[+] (hydrogen uptake positive), could metabolize the hydrogen synthesized by their nitrogenase. This H_2 oxidation supported the *ex planta* bacteroids nitrogenase activity and was coupled to ATP generation (Emerich et al., 1979) although not in all hup[+] *Rhizobium* (Nelson, 1979). To evaluate the benefit of the hydrogenase in the symbiosis, the growth of N_2 fed soybean was compared for a Hup[+] inoculant and its Hup[-] mutant. This growth was the highest with the Hup[-] strain on hydroponcally grown soybean (Drevon et al.,1987 b). Thus the N_2 fixation was lower in the presence of hydrogenase although less carbon substrate as consumed (Drevon et al., 1982). This result was understandable with the O_2 profile described above : since 02 was limiting

46

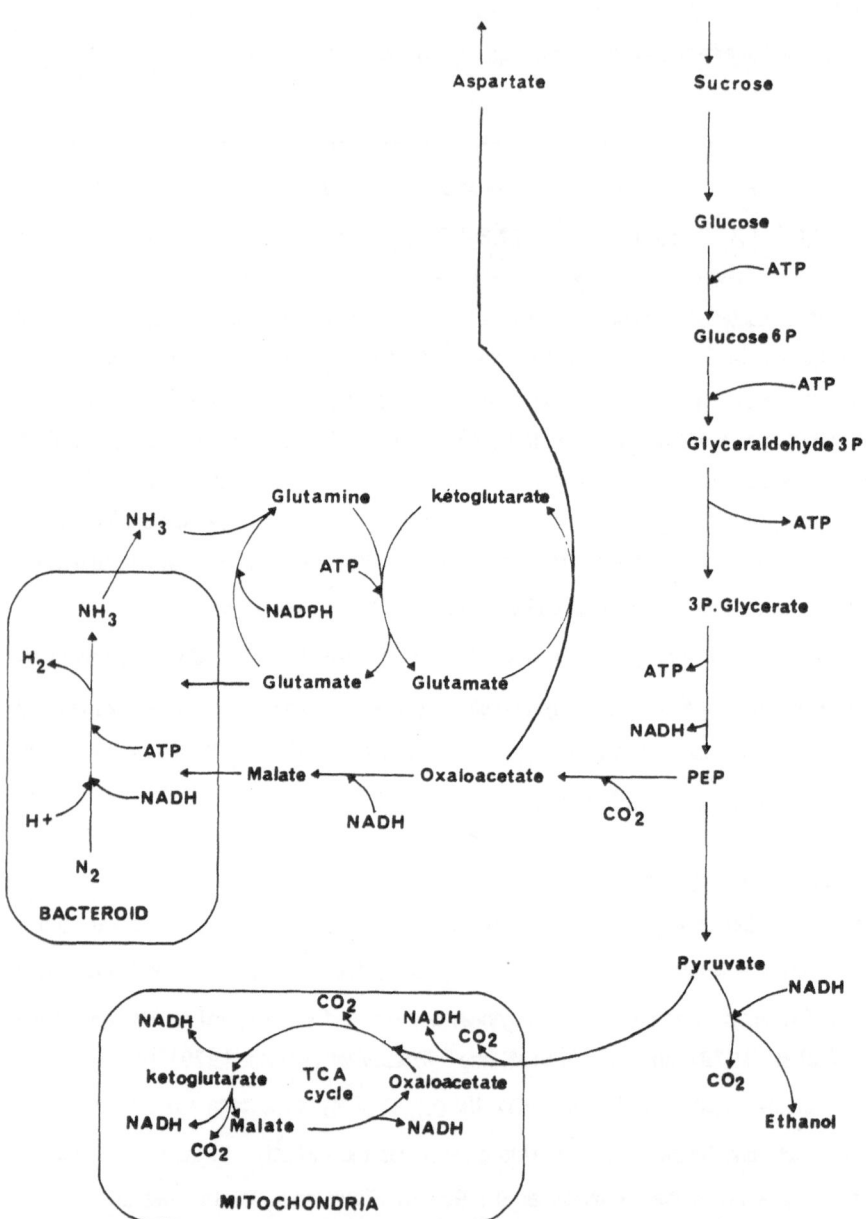

Fig.3 Hypothetical scheme of carbon and nitrogen metabolism in root-legume nodule

47

nitrogenase activity, the utilization of H_2 in Hup$^+$ strains would be beneficia' only if it were more efficient than the other bacteroid sources of reducing power. This would not be the case since the ARA of Hup $^+$ *ex planta* bacteroids at suboptimal O_2 tension was lower in the presence of H_2 than when it depended on endogenous substrates (Emerich et al., 1979; Godfroy et al., unpublished data). Also in pea, there was no beneficial effect of the hydrogenase (Cunningham et al., 1985).

Moreover it was repeatedly observed that the host-plant repressed the bacterial hydrogenase activity in some legumes species (Dixon et al., 1972 ; Lopez et al., 1983 ; Drevon et al., 1983-1987 b).This phenomenon was attributed to the shoots genotype (Bedmar and Phillips, 1984) and cyclic AMP could be implicated in this type of catabolite repression as a mediator (Lim and Shanmungan, 1979).

RELATIVE EFFICIENCY OF SYMBIOTIC FIXATION

Another way to limit the amount of energy lost in H_2 evolution would be to minimize the H_2 production by nitrogenase. The purified enzymatic complex produced at least one molecule of H_2 per molecule of N_2 reduced (Simpson and Burris, 1984), but more H_2 was synthesized when the turn over rate of the complex was decreased by lowering the ATP or the Fe protein concentration (Hageman and Burris, 1978).

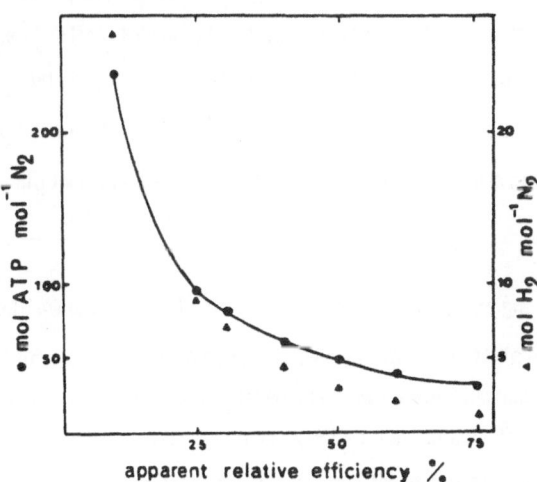

Fig. 4 Influence of the apparent relative efficiency on the theoretical cost of the reduction of one molecule of N_2

In the legume symbiosis, the amount of H_2 produced per N_2 reduced was estimated with Hup$^-$ strains by the measurement of the apparent relative efficiency proposed by Schubert and Evans (1976):

$$a\,RE = 1 - (\text{ rate of } H_2 \text{ evolution in air/ rate of } C_2H_4 \text{ evolution in } 10\% \ C_2H_2)$$

The figure 4 illustrates how the aRE is theoretically related to the cost of N_2 fixation based on the assumption that 2 ATP are hydrolysed for each electron transferred by the nitrogenase, and that one NADPH is equivalent to 4 ATP (Salsac et al., 1984). For example an aRE of 0,50 corresponds to 50% electrons allocated to protons reduction, thus to the following reaction of nitrogenase :

$$N_2 + 12H^+ + 12e^- \longrightarrow 2\,NH_3 + 3\,H_2$$

The reduction of one molecule of N_2 would require the transfer of 12 electrons, therefore the hydrolysis of 24 ATP and the oxidationn of 6 NADPH equivalent to 24 ATP ; so the total cost of N_2 reduction is 48 ATP.

On excised nodulated roots the aRE was generally lower than 0.75 (see Drevon and Salsac, 1984), which was also shown recently in the open flow in situ assay using a thermal conductivity detector for H_2 detection (Ollat et al., unpublished data). The aRE varied during the growth cycle and with the soybean cultivar from 0.40 to 0.65 (Drevon et al., 1986). It was increased by raising the external O_2 tension on excised nodulated roots (Drevon et al., 1982), and was the highest in the cultivar having the lowest critical oxygen pressure (Drevon et al., 1987 d) a parameter related to the metabolism limitation by O_2 (Saglio et al., 1984). This suggested that the host-plant would affect the aRE by its interaction with the O_2 supply of the bacteroids an assumption that has still to be checked on intact plants.

CONCLUSION

In the symbiosis having an optimal nitrogenase activity for external O_2 higher than 20%, the energy supply by the plant photosynthesis would not be a limiting factor of their activity in air Therefore an extra source of energy would be beneficial only if it is more efficient than the endogenous substrates ; this would not be the case for hydrogen which explains why in those symbiosis, the hydrogenase is not beneficial. In contrast, the reduction of H_2 production by nitrogenase, an ATP consuming process, should directly increase the amount of N_2 reduced.

These observations suggest that breeding for legumes and *Rhizobium* having their optimal symbiotic activity at pO_2 close to the ones in the soils, could be a possibility to improve the symbiotic-nitrogen fixation in its natural habitat.

REFERENCES

Béthenod, O., Prioul, J.L. and Deroche, M.E. 1984. Short-term inhibition of acetylene reduction activity by low CO_2 concentration around the nodulated roots of various legumes. Physiol. Vég., 22, 565-570.

Bedmar, E.J., and Phillips, D.A. 1984. A transmissible plant shoot factor promotes uptake hydrogenase activity in *Rhizobium* symbionts. Plant Physiol., 75, 629-633.

Coker, G.T., and Schubert, K.R. 1981. Carbon dioxide fixation in soybean roots and nodules. Plant Physiol., 67, 691-696.

Criswell, J.G., Havelka, U.D., Quebedeaux, B. and Hardy R.W.F. 1977. Effect of rhizosphere pO_2 on nitrogen fixation by excised and intact nodulated soybean roots. Crop Sc., 17, 39-44.

Cunningham, S.D., Kapulnik, Y., Brewin, N.J. and Phillips, D.A. 1985. Uptake hydrogenase activity determined by plasmid pRL6J1 in *Rhizobium leguminosarum* does not increase symbiotic nitrogen fixation. Appl. Environ. Microbiol., 50, 791-794.

Dixon, R.O.D. 1972. Hydrogenase in legume root nodules bacteroids : occurence and properties. Arch. Mikrobiol., 85, 193-201.

Drevon, J.J., Frazier, L., Russell, S.A. and Evans, H.J. 1982. Respiratory and nitrogenase activity of soybean nodules formed by hydrogen uptake negative mutant and revertant strain of *Rhizobium japonicum*. Plant Physiol., 70, 1341-1346.

Drevon, J.J., Tillard, P. and Salsac, L. 1983. Influence de *Vigna radiata* et *Vigna unguiculata* sur l'activité hydrogénase de la souche CB 756 de *Rhizobium* du groupe "cowpea", C.R. Acad. Sc., 296, 979-983.

Drevon, J.J. and Salsac, L. 1984. Relations entre le métabolisme de l'hydrogène et la fixation d'azote par les nodosités de légumineuses. Physiol. Vég., 22, 263-275.

Drevon, J.J., Tillard P. and Salsac, L. 1986. Variations of the apparent relative efficiency of nitrogen fixation by soybean root-nodules. Physiol. Vég., 24, 339-346.

Drevon, J.J. Kalia V.C., Heckmann, M.O. and Pedelahore, P. 1987 a. *In situ* open-flow assay of acetylene reduction activity by soybean root-nodules : influence of acetylene and oxygen. Plant Physiol. Biochem., in press.

Drevon, J.J., Kalia, V.C., Heckmann, M.O. and Salsac, L. 1987 b. Influence of the *Bradyrhizobium japonicum* hydrogenase on the growth of *Glycine* and *Vigna* species. Appl. Environ. Microbiol., 53, 610-612.

Drevon, J.J., Gaudillère, J.P., Bernoud, J.P., Jardinet, F. and Euvrard M. 1987 c. Influence of photon flux density and fluctuation on the nitrogen fixing *Glycine max* (L.Merr.). *Bradyrhizobium japonicum* symbiosis, Planta. in press.

Drevon, J.J., Kalia, V.C., Heckmann, M.O.., Tillard, P., Kimou, A., Pedelahore, P. and Salsac, L. 1987 d. Metabolisme de l'hydrogène et énergétique de la fixation symbiotique de l'azote chez le soja. In Les Colloques de l'INRA, 37, 55-66.

Eisbrenner, G. and Evans, H.J. 1983. Aspects of hydrogen metabolism in nitrogen-fixing legumes and other plant-microbe interactions. Ann. Rev. Plant Physiol., 34, 105-136.

Emerich, D.W., Ruiz-Argueso, T., Ching, T.M. and Evans, H.J. 1979. Hydrogen-dependent nitrogenase activity and ATP formation in *Rhizobium japonicum* bacteroids. J.Bacteriol., 137, 153-160.

Finn, G.A. and Brun W.A., 1982. Effect of atmospheric CO_2 enrichment on growth, structural carbohydrate content and root nodule activity in soybean. Plant Physiol., 69, 327-331.

Hageman, R.V. and Burris, R.H. 1980. Electron allocation to alternative substrates of *Azobacter vinelandii* is controlled by the electron flux through nitrogenase. Biochim. Biophys. Acta, 591, 63-75.

Hardy, R. W.F., Holsten, R.D., Jackson, E.K. and Burns, R.C. 1968. The acetylene-ethylene assay for N_2 fixation : laboratory and field evaluation. Plant Physiol., 43, 1185-1207.

Lim, S.T. and Shanmungam K.T. 1979. Regulation of hydrogen utilisation in *Rhizobium japonicum* by cyclic AMP. Biochim. Biophys. Acta, 584, 479-492.

Lopez, M.Carbonero, E., Cabrera, E. and Ruiz-Argueso, T. 1983. Effect of host on the expression of the H_2 uptake hydrogenase of *Rhizobium* in legume nodules. Plant Sc. Lett., 29, 191 -199.

Minchin, F.R., Witty, J.W., Sheehy, J.E. and Muller, M. 1983. A major error in the acetylene reduction assay : decreases in nodular nitrogenase activity under assay conditions. J.Exp. Bot., 34, 641-649.

Nelson, L.M., 1979. Hydrogen recycling by *Rhizobium leguminosarum* isolates and growth and nitrogen contents of pea plants (*Pisum sativum* L.) Appl. Environ. Microbiol., 45, 856-861.

Ralston, E.J. and Imsande, J. 1982. Entry of oxygen and nitrogen into intact soybean nodules. J.Exp. Bot., 33, 208-214.

Saglio, P.H., Rancillac, M. Bruzan F. and Pradet, A. 1984. Critical oxygen pressure for growth and respiration of excised and intact roots. Plant Physiol., 76, 151-154.

Salsac, L., Drevon, J.J., Zengbé, M., Cleyet-Marel, J.C. and Obaton , M.1984. Energy requirement of symbiotic nitrogen fixation. Physiol. Vég., 22, 509-521.

Schubert, K.R. and Evans, H.J. 1976. Hydrogen evolution : a major factor affecting the efficiency of nitrogen fixation in nodulated symbionts. Proc. Nat. Acad. Sci. USA, 73, 1207-1211.

Sheehy, J.E. , Minchin, F.R. and Witty, J.F.1985. Control of nitrogen fixation in a legume nodule : an analysis of the role of oxygen diffusion in relation to nodule structure. Ann. Bot., 55, 549-562.

Simpson, F.B. and Burris, R.H. 1984. A nitrogen pressure of 50 atmospheres does not prevent evolution of hydrogen by nitrogenase. Science, 224, 1095-1097.

Vance,C.P.,Stade,S. and Maxwell,C.A. 1983. Alfalfa root nodule carbon dioxide fixation and incorporation into amino acids. Plant Physiol.,72,469-473.

Weisz,P.R. and Sinclair,T.R. 1987.Regulation of Soybean nitrogen fixation in response to rhizosphere oxygen. Plant Physiol.,84,900-910.

RHIZOBIUM STRAIN EFFECT ON NITROGEN ACCUMULATION IN PEA RELATES TO PEP CARBOXYLASE ACTIVITY IN NODULES AND ASPARAGINE IN ROOT BLEEDING SAP

L. Rosendahl

Agricultural Research Department
Risø National Laboratory
DK-4000 Roskilde, Denmark

ABSTRACT

Pisum sativum L. cv. 'Bodil' was infected with selected strains of Rhizobium leguminosarum (R501, 128c53, B155, 18a or 1044). The Rhizobium strains influenced the activity of PEP carboxylase in the plant cytosol of the nodules and the composition of amino acids transporting the fixed N in the root bleeding sap. The specific activity of nodule PEP carboxylase was highest in the symbioses, which fixed the most N (B155, 18a and 1044). The root bleeding sap of these symbioses was composed of more asparagine and less glutamine than that of the two less effective symbioses (R501 and 128c53). The N yield of the symbioses was related neither to the nitrogenase-linked nodule respiration nor to the respiratory CO_2 efflux arising from growth and maintenance of the root system. The effect of the Rhizobium strains on N yield was maintained at maturity and reflected in seed yields.

Abbreviations - ARA, Acetylene reduction activity; PEP, phosphoenolpyruvate.

INTRODUCTION

Symbiotic N_2 fixation in legumes involves complex metabolic interactions between the cytoplasm of the infected host plant cell and the Rhizobium bacteroids. The Rhizobium genotype affects the N yield of the host plant (Rosendahl, 1984; Skøt et al., 1986). However, the physiological events, which underlies these differences has yet to be identified.

Nodule CO_2 fixation by PEP carboxylase activity is affected by the Rhizobium strain (Vance et al., 1983) and it has been demonstrated that nodule PEP carboxylase activity is positively correlated with ARA during nodule development (Christeller et al., 1977). The formation of C_4-dicarboxylic acids by PEP carboxylase activity in the nodule is an anaplerotic pathway, which may increase the efficiency of C u-

51

F. O'Gara et al. (eds.), Physiological Limitations and the Genetic Improvement of Symbiotic Nitrogen Fixation, 51–55.
© 1988 by Kluwer Academic Publishers.

tilization by providing C skeletons for assimilation of the
fixed N_2 in amide exporting legumes (Vance et al., 1983) and
by supplying the bacteroids with substrates for their metabolism (King et al., 1986). The pea plant assimilates the fixed
N_2 mainly as amides, and the amide composition of the root
bleeding sap is affected by the strain of Rhizobium (Rosendahl, 1984). Furthermore, the Rhizobium strain affects the
symbiotic C costs of N_2 fixation (Skøt et al., 1986; Witty et
al., 1983).

The aim of the present study was to evaluate if any of
the listed Rhizobium strain effects were related to N yield
of the symbiosis.

MATERIALS AND METHODS

Peas (Pisum sativum L. cv. 'Bodil') were grown in 1 litre
pots in vermiculite under Rhizobium controlled conditions in
a growth chamber (16/8 h light/dark, 19/12°C, 600 µmol m^{-2}
sec^{-1} PAR). At flowering, five weeks after seedling emergence, the assays were performed each on four replicate pots.
One set of four replicate pots was maintained until maturity.

The assays included: Kjeldahl N of tissues, PEP carboxylase activity in nodule cell-free extracts, soluble protein
in nodules, root bleeding sap collection and analysis of
amino acid content and gas exchange analysis on undisturbed
root systems in an open flow-through gas system. Further details on the techniques are reported by Rosendahl and Jakob-
sen (in press).

RESULTS AND DISCUSSION

The N accumulation at flowering of the two less effective
symbioses (R501 and 128c53) was on average 44% of the average
accumulated in the more effective symbioses (B155, 18a and
1044) (Table 1). This effect of the Rhizobium genotype on N
yield was also reflected in the yield of seeds at maturity
when the seeds of the R501 and 128c53 symbioses contained on
average 926 mg N pot^{-1}, whereas the three more effective symbioses contained on average 1563 mg N pot^{-1}. However, there
was no significant effect of the Rhizobium strain on the N

content in haulm and roots at maturity. The maintenance of Rhizobium strain effect until maturity and reflection in N yield of seeds is an important point, which is often neglected in physiological studies, despite of the fact that the ultimate goal in research on effectiveness of grain legumes is to increase seed yields.

TABLE 1 Nitrogen content of whole plants, nitrogenase- and PEP carboxylase activity of nodules and relative content of Asn and Gln in the root bleeding sap of Pisum sativum in symbiosis with various strain of Rhizobium leguminosarum. Values are mean of four replicate pots (two plants). Means within columns sharing the same letter are not significantly different at $P \leq 0.05$.

Rhizobium strain	Nitrogen content, mg pot^{-1}	ARA, μmol (g nodule DW)$^{-1}$ min^{-1}	PEP carboxylase, μmol (mg protein)$^{-1}$ min^{-1}	Asn, % of total amino acids	Gln, % of total amino acids
R501	110^a	4.1^a	0.7^a	56^b	29^c
128c53	211^b	2.9^a	0.7^a	34^a	43^d
B155	378^c	5.4^{ab}	1.2^b	62^c	18^b
18a	356^c	9.1^c	1.2^b	62^c	12^{ab}
1044	354^c	8.1^{bc}	1.1^b	67^c	9^a

The PEP carboxylase activity in the nodules of the low-yielding symbioses constituted on average 59% of that recorded in the high-yielding symbioses (Table 1). The earlier studies by Vance et al. (1983) on Rhizobium strain effects on PEP carboxylase activity in legume nodules involved a comparison of a Rhizobium strain, which formed an active nitrogenase as opposed to strains lacking nitrogenase activity in the symbiosis. This work supplements the results by Vance et al. (1983) as the investigated symbioses all posses an active nitrogenase (Table 1), and the data indicate a positive relation between N yield of the symbioses and PEP carboxylase activity in the nodules.

The oxaloacetate, produced by PEP carboxylase activity in the nodules, provides a source of C skeletons for assimilation of the symbiotically fixed N_2 in pea, which is an amide-

transporting legume (Sprent, 1980; Vance et al., 1983). The
main pathway of N assimilation in the plant cytosol of the
nodule proceeds via glutamine synthetase, glutamate synthase,
aspartate amino transferase and asparagine synthetase. The C
skeletons for aspartate and asparagine are derived from oxaloacetate. Accordingly, the symbioses with the higher PEP
carboxylase activity (B155, 18a and 1044) had the highest
relative content of asparagine in the root bleeding sap (Table 1). Glutamine was more predominant in the root bleeding
sap of the two least effective symbioses, R501 and 128c53
(Table 1). However, from the present work it cannot be established whether this glutamine originates from a short cut in
the N assimilation pathway.

The nitrogenase-linked nodule respiration (mol CO_2/mol
C_2H_4) was unaffected by the investigated strains of <u>Rhizobium</u>
(Table 2). The variation in CO_2 efflux from growth and maintenance reflected reasonably well the variation in total below ground biomass (Table 2). Neither of these parameters related to N yield of the symbiosis.

TABLE 2   Nitrogenase-linked nodule respiration and respiratory CO_2 efflux from growth and maintenance in root
systems of <u>Pisum sativum</u> in symbiosis with various
strains of <u>Rhizobium leguminosarum</u>. Values are mean of
four replicate pots (two plants). Means within columns
sharing the same letter are not significantly different
at P ≤ 0.05.

Rhizobium strain	Carbon cost mol CO_2 (mol C_2H_4)$^{-1}$	CO_2 efflux from growth and maintenance µmol pot^{-1} min^{-1}
R501	2.0[a]	1.1[a]
128c53	1.8[a]	2.6[b]
B155	1.7[a]	2.9[b]
18a	2.1[a]	2.6[b]
1044	2.0[a]	2.7[b]

The present results demonstrate that the <u>Rhizobium</u> strain
determined N accumulation of the host plant is related to

nodule PEP carboxylase activity and to the composition of the
N solutes exported to the top of the investigated symbioses.
However, the causal relationship between the N_2 fixation and
PEP carboxylase activity cannot be deduced from this work.
The mechanisms by which the Rhizobium strain affects the activity of plant enzymes, which are involved in an optimal
function of the symbiosis, still need to be elucidated. The
microenvironment within the effective nodule differs markedly
from that of an ineffective nodule or uninfected root tissue.
This may affect the activity of plant enzymes within the nodule (Larsen and Jochimsen, 1986). Investigations have been
initiated to elucidate if microenvironmental conditions characteristic for effective nodules may stimulate PEP carboxylase activity.

REFERENCES

Christeller, J.T., Laing, W.A. and Sutton, W.D. 1977. Carbon
    dioxide fixation by lupin root nodules. I. Characteriza-
    tion, association with phosphoenolpyruvate carboxylase,
    and correlation with nitrogen fixation during nodule de-
    velopment. Plant Physiol., 60, 47-50.
King, B.J., Layzell, D.B. and Canvin, D.T. 1986. The role of
    dark carbon dioxide fixation in root nodules of soybean.
    Plant Physiol., 81, 200-205.
Larsen, K. and Jochimsen, B.U. 1986. Expression of nodule-
    specific uricase in soybean callus tissue is regulated by
    oxygen. EMBO Jour., 5(1), 15-19.
Rosendahl, L. 1984. Rhizobium strain effects on yield and
    bleeding sap amino compounds in Pisum sativum. Physiol.
    Plantarum, 60, 215-220.
Skøt, L., Hirsch, P.R. and Witty, J.F. 1986. Genetic factors
    in Rhizobium affecting the symbiotic carbon costs of N_2
    fixation and host plant biomass production. J. Appl.
    Bacteriol., 61, 239-246.
Sprent, J.I. 1980. Root nodule anatomy, type of export pro-
    duct and evolutionary origin of some Leguminosae. Plant
    Cell Environ., 3, 35-43.
Vance, C.P., Stade, S. and Maxwell, C.A. 1983. Alfalfa root
    nodule carbon dioxide fixation. I. Association with ni-
    trogen fixation and incorporation into amino acids. Plant
    Physiol., 72, 469-473.
Witty, J.F., Minchin, F.R. and Sheehy, J.E. 1983. Carbon
    costs of nitrogenase activity in legume root nodules
    determined using acetylene and oxygen. J. Exp. Bot., 34,
    951-963.

FLAVONOID - RHIZOBIUM INTERACTIONS
IN LOTUS SPECIES

J.R. Rao, J.E. Cooper
Department of Agricultural and Food Bacteriology
The Queen's University of Belfast, Belfast BT9 5PX
Northern Ireland, UK

ABSTRACT

Flavonoid compounds are present in the roots and shoots of Lotus species, either in glycosidic combination or as flavolans which are usually polymerised to form condensed tannins. Some strains of Lotus rhizobia are highly sensitive to flavolans in vitro and it has been suggested that nitrogen fixation by such strains is reduced or inhibited. This paper examines the relationship between symbiotic effectiveness and flavolan sensitivity for a large number of rhizobial isolates from Lotus corniculatus and Lotus pedunculatus. The formation of flavolans from their flavonoid precursors in roots and nodules was also studied. No correlation was found between flavolan sensitivity and symbiotic effectiveness and some strains were highly effective on a host while exhibiting sensitivity to its root flavolans. Polymeric flavolans were detected in the roots and ineffective nodules of plants but not in effective nodules. Rhizobia would not encounter polymeric flavolans in the roots of Lotus at the stages of infection thread formation and early nodule development. Effective rhizobia may interact with flavolan precursors in nodule tissues, preventing their further metabolism by the plant or condensation to tannins. A mechanism is proposed for entry of flavolan precursors into the TCA cycle and their utilization as energy sources by effective rhizobia.

INTRODUCTION

Tannins are water soluble polymers of flavolans which are present in high concentrations in the roots and shoots of Lotus spp. The synthesis of these compounds from monomeric flavonoids in plant tissues is outlined in Fig. 1. Tannins are important secondary metabolites for several reasons: they interact with proteins, enzymes and certain polysaccharides (Kefeli and Kutacek, 1977) and their presence in forage legumes is considered desirable due to their ability to precipitate proteins (Swain and Bate-Smith, 1962), thereby minimising formation of the proteinaceous foam which causes pasture bloat in ruminant animals (Goplen et al., 1980).

With regard to nitrogen fixation, however, accumulations of polymeric flavolans in root and nodule tissues of Lotus may be toxic to Rhizobium (or Bradyrhizobium). Pankhurst et al. (1979) observed that the flavolan

F. O'Gara et al. (eds.), Physiological Limitations and the Genetic Improvement of Symbiotic Nitrogen Fixation, 57–64.
© 1988 by Kluwer Academic Publishers.

58

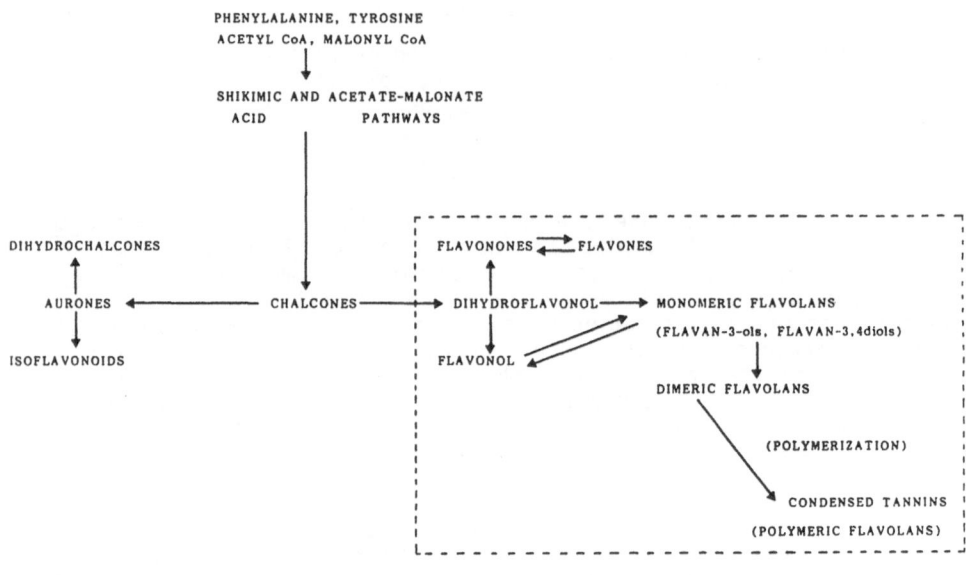

Fig. 1   General pathway of flavolan biosynthesis in plant tissues (modified from Vickery and Vickery, 1981)

content of Lotus roots was greater in plants which were ineffectively nodulated.   Pankhurst and Jones (1979) also showed that an ineffective R. loti strain was more sensitive than an effective strain to flavolans extracted from the common host plant.   Flavolan tolerance was associated with a strain's ability to bind more of the compound in the exponential growth phase (Pankhurst et al., 1982).   Since a strain of R. loti may be ineffective on one Lotus spp. and effective on another, and since flavolan composition is species dependent, it has been suggested that ineffectiveness may be a consequence of a strain's sensitivity to a particular type of plant flavolan during the early stages of infection thread formation and growth (Pankhurst et al., 1982).

In this paper we examine the relationship between symbiotic effectiveness and in vitro flavolan sensitivity for a wide range of rhizobia isolated mainly from L. pedunculatus and L. corniculatus.   We also present data on the distribution of flavolans and their flavonoid precursors in roots and nodules of effectively and ineffectively nodulated Lotus spp.

MATERIALS AND METHODS

Strain effectiveness

Effectiveness of 24 fast-growing (R. loti) and 18 slow-growing (Bradyrhizobium) strains on L. pedunculatus and L. corniculatus (10 plants per strain) was determined in N-free rooting solution. Dry weight of plant shoots was measured after eight weeks' growth at 25° C with a 17 h day length. Effectiveness was scored on a scale from 0 (ineffective) to 4 (highly effective).

Extraction of flavolans

Flavolans were extracted from the roots of L. pedunculatus and L. corniculatus in acetone-water (1:1) solution followed by separation of other plant phenolics using diethyl ether and ethyl acetate (Broadhurst and Jones, 1978). The final aqueous phase containing flavolans was further purified by the method of Jones et al. (1976) and concentration was estimated by colorimetric reaction with vanillin-HCl.

Sensitivity of rhizobia to flavolans

This was assessed by monitoring viable counts of strains in mannitol-based medium supplemented with purified root flavolans (25 μg ml^{-1}) from L. pedunculatus or L. corniculatus. Comparison was made with uninhibited growth in non-supplemented medium.

Distribution of flavonoids in roots and nodules

Methanol extracts of roots and nodules (effective and ineffective) of L. pedunculatus and L. corniculatus were subjected to 2 D paper chromatography using (i) BAW (butanol : acetic acid : water; 4 : 1:5; 17 h) and (ii) Forestal (acetic acid : water : HCl; 30 : 10 : 3; 15 h) solvent systems. Component flavonoids were identified by colour reactions under UV light in the presence or absence of ammonia and after treatment with high sensitivity spray reagents as described by Markham (1982). Concentrations of flavonoids were assessed by comparing absorbance values of ethanol extracts of chromatograph spots with co-chromatographed flavonoids of known concentration at appropriate wavelengths in a spectrophotometer (Mabry et al., 1970).

RESULTS AND DISCUSSION

Fig. 2 shows the relationship between symbiotic effectiveness and flavolan content of L. pedunculatus roots for 40 Lotus rhizobia comprising both R. loti and Bradyrhizobium strains. Plants nodulated by highly effective rhizobia did not contain lower amounts of flavolan in their roots and concentrations of these compounds found in all mature plants would be toxic to rhizobia in vitro.

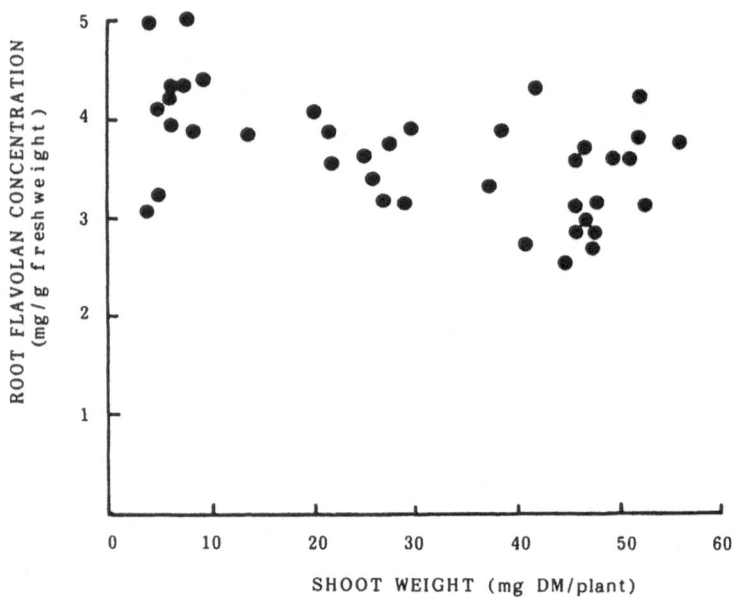

Fig. 2 Relationship between root flavolan content of Lotus pedunculatus and symbiotic effectiveness for forty strains of Lotus rhizobia.

In sensitivity tests on a selection of strains flavolans from L. pedunculatus and L. corniculatus caused significant inhibition of rhizobial growth at a concentration of 25 µg·ml^{-1} (Table 1). The least sensitive strain was capable of 32% of its multiplication in non-supplemented growth medium. Some strains (LP13, LP18, LC18) were completely inhibited by this concentration of flavolan and there was no evidence to suggest that effective strains on either host were less sensitive to flavolans in their growth medium. Indeed, some strains were highly effective on a host while exhibiting a high degree of in vitro sensitivity to flavolans.

TABLE 1 Relationship between in vitro flavolan sensitivity of rhizobia and their symbiotic effectiveness on Lotus pedunculatus and Lotus corniculatus.

Strain	Original host	% inhibition of growth in media supplemented with root flavolans extracted from:		Effectiveness of N fixation (scale 0-4, ineffective-effective) on:	
		L. pedunculatus	L. corniculatus	L. pedunculatus	L. corniculatus
LP1	L. pedunculatus	94	90	4	0
LP2	"	90	78	4	0
LP13	"	100	79	4	0
LP18	"	99	84	4	1
CC814s	L. hispidus	90	90	4	0
LP24	L. pedunculatus	73	88	3	3
LC3	L. corniculatus	66	80	0	3
LC5	"	85	94	0	3
LC16	"	68	97	0	3
LC18	"	98	92	1	3
LC23	"	77	83	0	2

Monomeric flavolans and flavonols (eg quercetin, kaempferol) were identified in developing effective and ineffective nodules but polymeric flavolans were only detected in ineffective nodules 15 days after first appearance (Fig. 3). The pattern of flavonoid distribution was similar in both Lotus spp. These data indicate that effective strains of Lotus rhizobia would not be exposed to condensed tannins during the period of nodule development which includes the onset of nitrogen fixation. Polymeric flavolans did not accumulate in mature, effective nodules (data not presented).

The results indicate that no simple relationship exists between the effectiveness of Lotus rhizobia and their sensitivity to host plant condensed tannins. Our findings do raise the possibility of an interaction between effective rhizobia and flavonoids at the stage of biosynthesis prior to formation of polymeric flavolans. Rhizobia are known to utilize a number of aromatic compounds, including flavonoids, as carbon and energy sources (Muthukumar et al., 1982; Parke and Ornston, 1984) and the production of ATP from their metabolism has been established through several metabolic pathways (Chen et al., 1984). Alcohols and aldehydes, present in the infected host cytosol and utilized by Rhizobium bacteroids, could be a contributory source of ATP for N fixation (Peterson and La Rue, 1981; Tajima and La Rue, 1982) and certain flavonoids are thought to contribute towards bacteroid metabolism in alfalfa (Vance, 1978). The glycosidic association of the simpler flavonoids also has

62

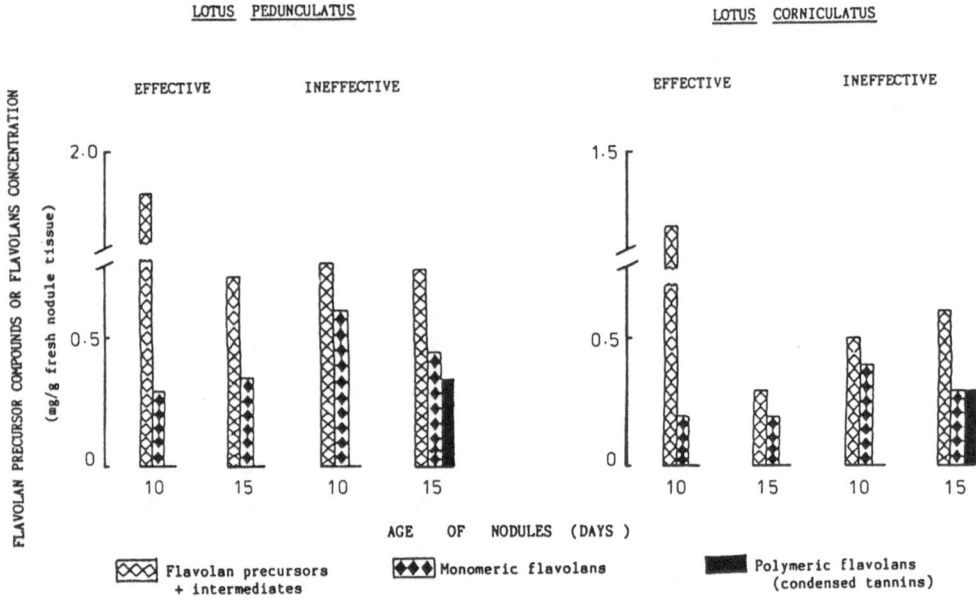

Fig. 3 Presence of flavolan precursor compounds, monomeric and polymeric flavolans in effective and ineffective root nodules of *Lotus pedunculatus* and *Lotus corniculatus*.

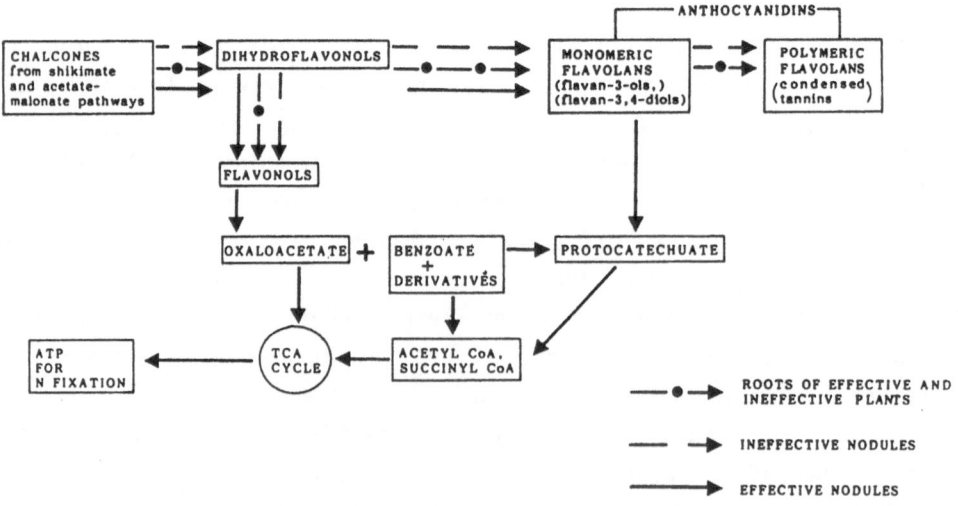

Fig. 4 Proposed mechanism for flavonoid metabolism in roots and nodules of *Lotus pedunculatus*.

relevance in this context, since mobility of the compound is increased as a consequence of its higher vacuolar solubility (McClure, 1975).

A proposed mechanism for an alternative fate of the precursors of polymeric flavolans in effective Lotus root nodules is shown in Fig. 4. In this scheme flavonols, flavan-3-ols and flavan-3,4-diols would be converted directly to oxaloacetate or to benzoate and protocatechuate which in turn yield succinyl CoA and acetyl CoA. These products could then enter the TCA cycle and generate additional ATP for N fixation. In ineffective nodules the same flavonoids would be expected to accumulate and finally form condensates of polymeric flavolans.

REFERENCES

Broadhurst, R.B. and Jones, W.T. 1978. Analysis of condensed tannins using acidified vanillin. J. Sci. Fd. Agric., 29, 788-794.

Chen, Y.P., Glenn, A.R. and Dilworth, M.J. 1984. Uptake and oxidation of aromatic substrates by Rhizobium leguminosarum MNF3841 and Rhizobium trifolii TAI. FEMS Microbiol. Lett., 21, 201-205.

Goplen, B.P., Howarth, R.E., Sarkar, S.K. and Lesins, K. 1980. A search for condensed tannins in annual and perennial species of Medicago, Trigonella and Onobrychis. Crop Sci., 20, 801-804.

Jones, W.T., Broadhurst, R.B. and Lyttleton, J. 1976. The condensed tannins of pasture legume species. Phytochemistry, 15, 1407-1409.

Kefeli, V.I. and Kutacek, M. 1977. Phenolic substances and their possible role in plant growth regulation. In "Plant Growth Regulation" (Ed. P.E. Pilet). (Springer, Berlin). pp. 181-188.

Mabry, T.J., Markham, K.R. and Thomas, M.B. 1970. The Systematic Identification of Flavonoids. (Springer, Berlin).

Markham, K.R. 1982. Techniques of Flavonoid Identification. (Academic Press, London).

McClure, J. 1975. Physiology and function of flavonoids. In "The Flavonoids" (Ed. J.B. Harborne, T.J. Mabry and H. Mabry). (Chapman and Hall, London). pp. 970-1055.

Muthukumar, G., Arunakumari, A. and Mahadevan, A. 1982. Degradation of aromatic compounds by Rhizobium spp. Pl. Soil, 69, 163-169.

Pankhurst, C.E., Craig, A.S. and Jones, W.T. 1979. Effectiveness of Lotus root nodules. I. Morphology and flavolan content of nodules formed on Lotus pedunculatus by fast-growing Lotus rhizobia. J. Exp. Bot., 30 1085-1093.

Pankurst, C.E. and Jones, W.T. 1979. Effectiveness of Lotus root nodules. II. Relationship between root nodule effectiveness and in vitro sensitivity of fast-growing Lotus rhizobia to flavolans. J. Exp. Bot., 30, 1095-1107

Pankhurst, C.E., Jones, W.T. and Craig, A.S. 1982. Bactericidal effect of Lotus pedunculatus root flavolan on fast-growing Lotus rhizobia. J. Gen. Microbiol., 128, 1567-1576.

Parke, D. and Ornston, L.N. 1984. Nutritional diversity of Rhizobiaceae revealed by auxanography. J. Gen. Microbiol., 130, 1743-1750.

Peterson, J.B. and La Rue, T.A. 1981. Utilization of aldehydes and alcohols by soybean bacteroids. Plant Physiol., 68, 489-493.

Swain, T. and Bate-Smith, E.C. 1962. Flavonoid compounds. In "Comparative Biochemistry", vol. 3 (Ed. A.M. Florkin and H.S. Mason), (Academic Press, New York). pp. 755-809.

Tajima, S. and LaRue, T.A. 1982. Enzymes for acetaldehyde and ethanol formation in legume nodules. Plant Physiol., 70, 388-392.

Vance, C.P. 1978. Comparative aspects of root and root nodule secondary metabolism in alfalfa. Phytochemistry, 17, 1889-1891.

Vickery, M.L. and Vickery, B. 1981. Secondary Plant Metabolism. (Macmillan, London).

PERIBACTEROID MEMBRANE STABILITY AND PHYTOALEXIN PRODUCTION IN LEGUME NODULES

A. Wolff, E. Morschel, C. Zimmermann, M. Parniske, S. Bassarab, R. Mellor and D. Werner

Fachbereich Biologie der Philipps-Universitat, Karl-von-Fischer-str., D-3550 Marburg/L, FRG.

ABSTRACT

The biogenesis and functions of the peribacteroid membrane (PBM) are summarised. Examples for the structural variability of the PBM and the peribacteroid space in Vicia faba are given. In order to use phytoalexin production in nodules as a biochemical marker for istabilities of the PBM in Vicia faba , the development of radio-immune-assays for the detection of wyerone and wyeronic acid is described. Thin layer chromatography and HPLC techniques were used for the purification of these phytoalexins. The nitrogenase activity of the nodules studied was determined from field grown plants as well as from plants infected with nod^+ fix^+ defined strains of Rhizobium leguminosarum.

INTRODUCTION

The faba bean (Vicia faba), a major seed legume with a world annual production of 4.3 million t and an EEC production of 0.44 million t (FAO data from 1981), has a stagnating or decreasing area of production. The low yield stability which is also accompanied by a large variation in nitrogen fixation is one of the major handicaps of the cultivars of this crop. Reported data of nitrogen fixation vary between 40 and 500 kg N per hectare per year. On the other hand the potential of this old European crop is rather high. Nodulation and nitrogen fixation in Vicia faba is distinguished by

- a high dominance of the endogenous strain of Rhizobium leguminosarum against inoculating strains;

- the large number of often very small nodules per plant (up to

65

F. O'Gara et al. (eds.), Physiological Limitations and the Genetic Improvement of Symbiotic Nitrogen Fixation, 65–74.
© 1988 by Kluwer Academic Publishers.

1000);

- vacuolated host cells after bacteroid differentiation;
- a sometimes low responsiveness to shading experiments;
- an often high responsiveness to water conditions and
- a low responsiveness to low and medium N-fertilisation.

MATERIALS AND METHODS

Growth and infection of Vicia faba cultivars under defined conditions was as described previously (Wolff et al., 1986).

Nitrogenase activity by the intact plant assay was essentially the same as described for other legumes (Werner et al., 1975).

Details of the purification of wyerone and wyeronic acid from cotyledons of Vicia faba infected with Botrytis cinerea are given in the legend to Fig. 7. To produce ^{14}C-labelled wyeronic acid, ^{14}C-acetate was added to the cotyledons before infection and absorbance of the fractions was monitored at 350 nm.

Purified wyeronic acid was bound to bovine serum albumin (BSA) by incubation with equimolar concentrations of N, N'-dicyclohexylcarbodiimide and N-hydroxysuccinimide after dissolution in acetone. The reaction was for 2 h at room temperature by which time the acetone evaporated. The sediment was redissolved in distilled water and an equimolar amount of BSA added. To separate bound from unbound wyeronic acid the samples were dialysed against water overnight. For immunisation 200 µg of BSA bound wyeronic acid and 150 µl Freud adjuvant were used.

Production of p-aminohippuric acid substituted BSA was done according to Weiler (1980).

The radio-immune-assays were carried out in Eppendorf vials (1.5 ml capacity). Each vial contained: 100 µl 0.05 M acetate buffer, pH 6.2; 100 µl 10% methanol; 100 µl of the corresponding diluted antiserum and 50 µl tracer with 5 pm ^{14}C-wyerone, with 900 dpm.

The fine structure analysis of nodules was performed as described previously (Werner et al., 1984).

RESULTS AND DISCUSSION

Nitrogen fixation capacity

The cultivars of Vicia faba used in this study show a typical response in their nitrogen fixation capacity to variations in the oxygen supply (Fig. 1) and infection by Rhizobium leguminosarum strains with different effeciency (Fig. 2). The maximum rate of 30 nmol C_2H_2 h^{-1} mg^{-1} nodule freash weight is relatively high compared to other nodules (Werner, 1987).

Fine structure variability of the peribacteroid membrane

By comparing different cultivars of Vicia faba at several stages of nodulation, we observed a number of variations in the stability of the peribacteroid membrane. Close to the large host cell nucleus (Fig.3) the peribacteroid membrane seems to be open or leaky. However, this observation is difficult to quantify and so far it can also not be decided if it is typical for the cultivar Kristall at certain stages of development. In the cultivar Minica we found in many cells very large peribacteroid membrane compartments with several bacteroids inside an apparently very large peribacteroid space (Fig. 4). In the cultivar Diana we noticed inside a large infection thread bacteroids which accu-

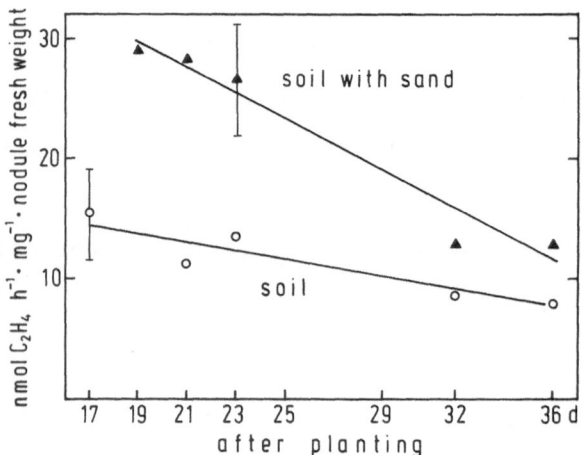

Fig. 1 Nitrogenase activity in <u>Vicia</u> <u>faba</u> cv Herra in soils from a bean field (o) and in the <u>same</u> <u>soil</u> mixed with sand (▲). Days after planting.

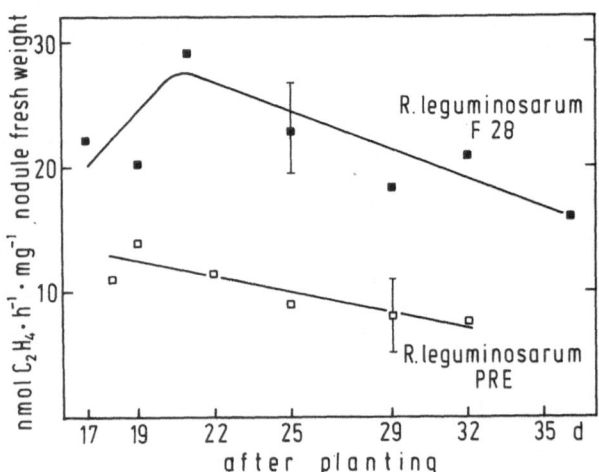

Fig. 2 Nitrogenase activity in <u>Vicia</u> <u>faba</u> cv Herra grown in Perlite, infected with <u>Rhizobium</u> <u>leguminosarum</u> PRE (□), infected with <u>Rhizobium</u> <u>leguminosarum</u> F 28 (■).

mulate poly-β-hydroxybutyrate (PHB) (Fig. 5). The regular arran-
gement of the PBM around the bacteroids is also shown in this
figure. The release of a bacteroid from the infection thread
with a stable peribacteroid membrane developing is shown in Fig.
6. It is very difficult to give quantitative data for unstable
or leaky peribacteroid membranes in many combinations of micro-
and macrosymbionts, and especially for several areas of the infected zone of undetermined nodules during development. The development of a radio-immune-assay for phytoalexins which can be
used as a biochemical marker indicating the various degrees of
instability of this symbiotic structure is therefore all the
more important.

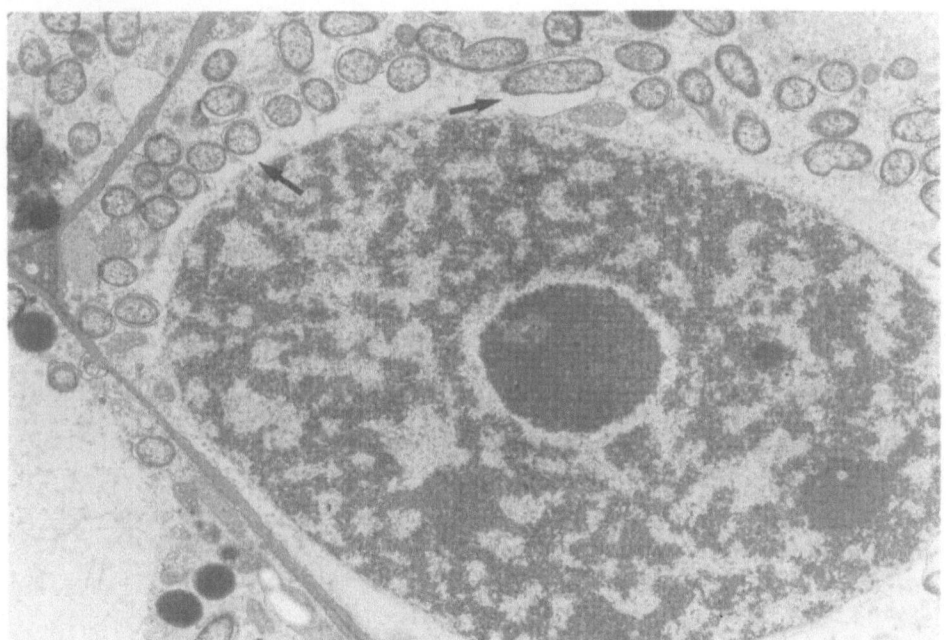

Fig. 3. Fine structure of an infected cell from nodules of Vicia
faba cv. Kristall. The arrows indicate changed peribacteroid
membranes neighbouring the host cell nucleus (7800x).The field
grown plants used in this study (Figs. 3-6) were kindly supplied
by Dr. Ebmeyer and Prof. Robbelen,Institut fur Pflanzenbau und
Pflanzenzuchtung, Universitat Gottingen.

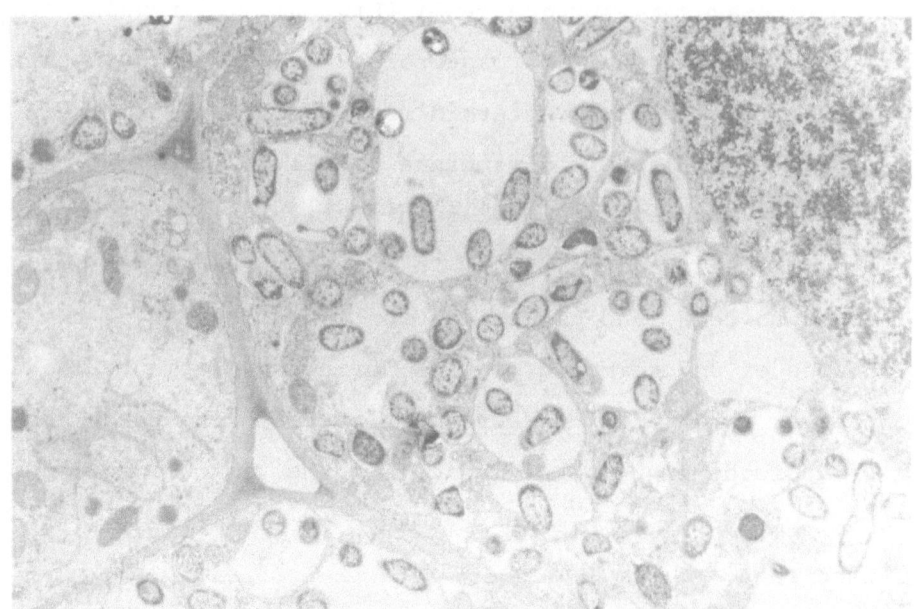

Fig. 4. Fine structure of infected cells from nodules of <u>Vicia faba</u> cv. Minica. Several bacteroids located in one peribacteroid membrane vesicle (7300x).

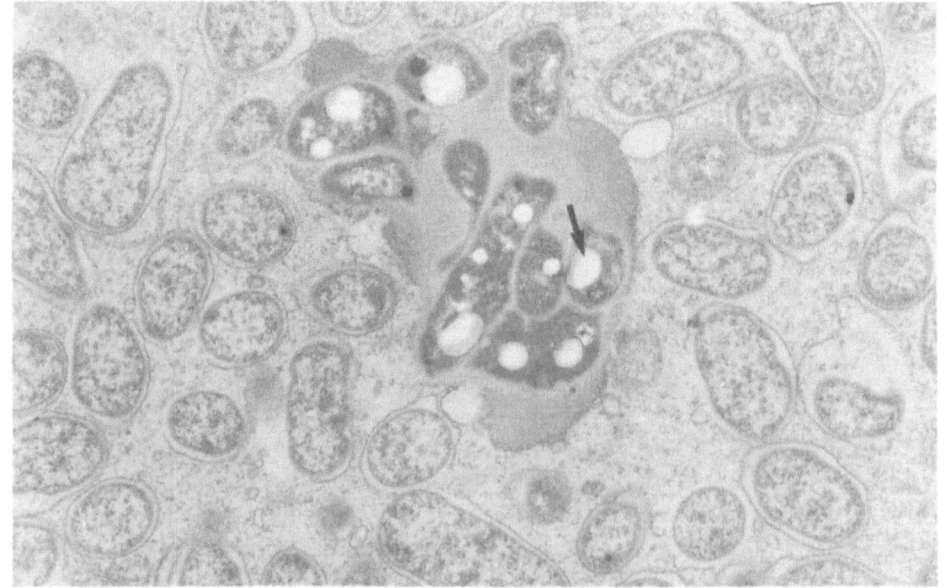

Fig. 5. Fine structure of an infected cell from nodules of <u>Vicia faba</u> cv. Diana. In the centre an infection thread with bacteroids accumulating poly-β-hydroxybutyrate (arrow) can be seen (15 700x).

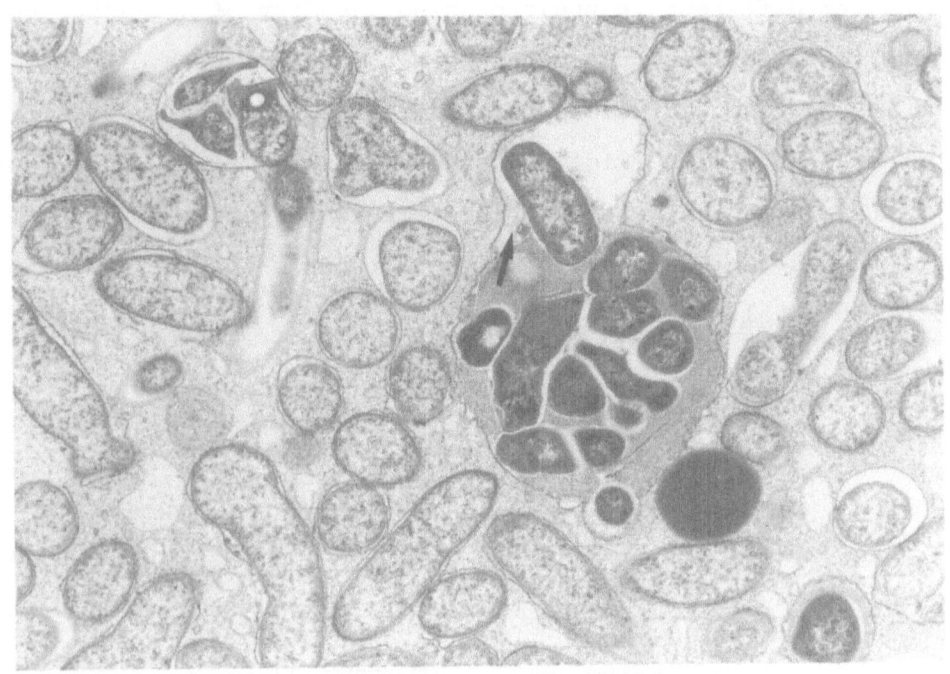

Fig. 6. Fine structure of an infected cell from nodules of Vicia
faba cv. Kristall showing in the centre (arrow) the release of a
bacteroid, enclosed by the developing peribacteroid membrane ,
from an infection thread into the host cytoplasm (15 700x).

Development of RIA-tests for phytoalexins in Vicia faba

The procedure used for the purification of wyerone and
wyeronic acid was an improvement over previous methods (Mansfield et al., 1980; Mansfield and Hargreaves, 1984; Wolff et
al., 1986) and is summarised in Fig.7. A combination of thin
layer chromatography and HPLC was used to obtain rather pure
fractions of wyerone and wyeronic acid. ^{14}C-labelled wyerone was
purified after incubation of broad bean seedlings with ^{14}C-
acetate (3.4×10^4 Bq/g seedling).

Wyeronic acid was coupled to bovine serum albumin by two
methods (Fig. 8), either directly or to an aminohippuric acid

BSA conjugate. The antibodies produced against this conjugate were rather different in their binding capacity of ^{14}C-labelled wyerone. The best results so far were obtained with a directly bound BSA wyeronic acid conjugate antiserum. The unspecific binding was about 20% (Fig. 9), and the maximum binding efficiency was 60%. Therefore this test so far has not the same efficiency as the RIA-test for glyceollin from soybean (Grisebach et al., 1986).

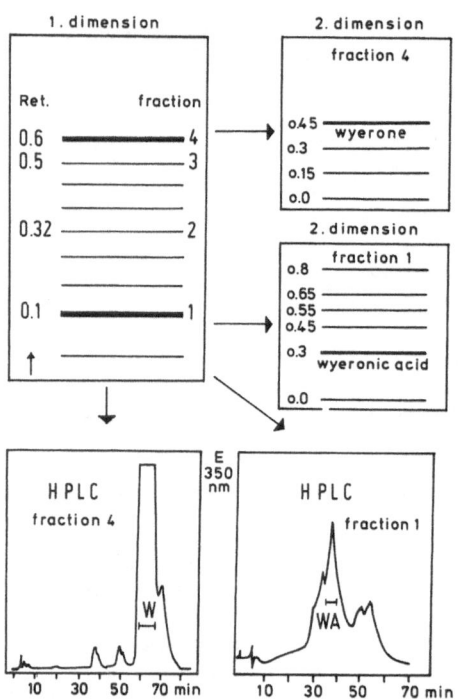

Fig.7. Purification of wyerone and wyeronic acid from cotyledons of Vicia faba cv. Kristall infected with Botrytis cinerea. For thin layer chromatography the mobile phase in the first dimension was hexane/acetone 2:1; the mobile phase in the second dimension for fraction 4 was hexane/acetone 4:1 and that for fraction 1 was ether/methanol 6:1. Wyerone and wyeronic acid were also purified after the first dimension by HPLC with a linear gradient from methanol/water 40:60 to methanol/water/acetonitrile 40:40:20. The flow rate was 0.8 ml/min.

Coupling of wyeronic acid to BSA for antibody production

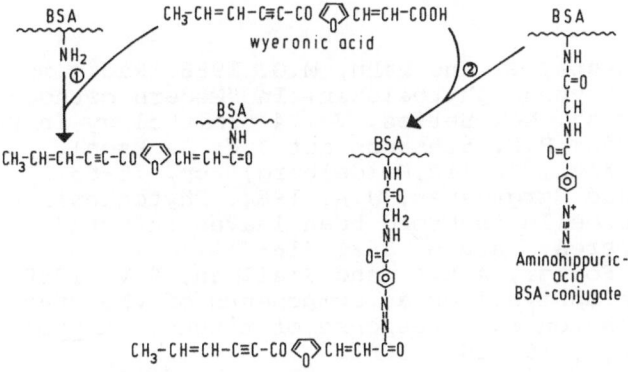

Fig. 8. Coupling of wyeronic acid to BSA for antibody production.

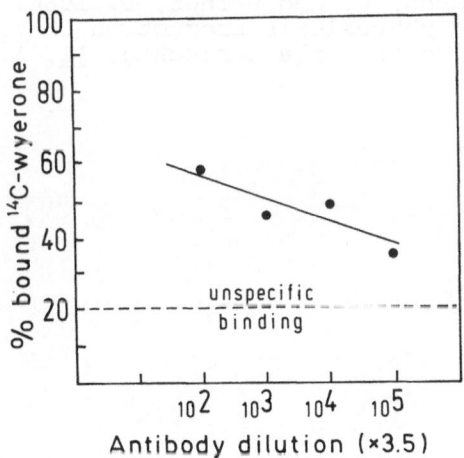

Fig. 9. RIA-tests for ^{14}C-labelled wyerone binding against antibody dilution 10^{-2}-10^{-5} fold.

74

ACKNOWLEDGEMENTS

    Support of the EC (Energy in Agriculture, Exchange of
Scientists) and the BMFT, project "Wechselwirkungen zwischen
Mikroorganismen and Kulturepflanzen" is gratefully acknow-
leged.

REFERENCES

Grisebach, H., Moesta, P. and Hahn, M.G. 1986. Radioimmuno-
    assay for a soybean phytoalexin. In "Modern methods of
    Plant Analysis", New Series, Vol.4 "Immunology in Plant
    Sciences" (Eds. H.F. Linskens anf J.F. Jackson).
    (Springer Verlag, Berlin,Heidelberg). pp. 75-85.
Mansfield, J.W. and Hargreaves, J.A. 1984. Phytoalexin produc-
    tion by live cells in broad bean leaves infected by
    Botrytis cinerea. Nature, 252, 316-317.
Mansfield, J.W., Porter, A.E.A. and Smallman, R.V. 1980. Di-
    hydrowyerone derivatives as components of the urano-
    acetylenic phytoalexin response of tissues of Vicia faba.
    Phytochemistry, 19, 1057-1061.
Weiler, E.W. 1980. Radioimmunoassays for the differential and
    direct analysis of free and conjugated abscisic acid in
    plant extracts. Planta, 148, 262-272.
Werner, D., Morschel, E., Kort, R., Mellor. R.B. and Bassarab,
    S. 1984. Lysis of bacteroids in the vicinity of the host
    cell nucleus in an ineffective (fix) root nodule of soy-
    bean (Glycine max). Planta, 162, 8-16.
Werner, D., Wilcockson, J. and Zimmermann, E. 1975. Adsorpt-
    ion and selection of rhizobia by ion exchange papers.
    Arch. Microbiol. 105, 27-32.
Wolff, A., Zimmermann, C. and werner, D. 1986. Nodule compart-
    mentation and phytoalexin production in Vicia faba and
    Glycine max. Vortr. Pflanzenzuchtg. 11, 174-185.

SECTION II : ENVIRONMENTAL FACTORS AFFECTING NITROGEN FIXATION

LIMITATIONS AND BENEFITS OF OXYGEN DIFFUSION CONTROL IN LEGUME NODULES

F. R. Minchin, J. F. Witty, L. Skøt

Department of Plant and Cell Biology, Welsh Plant Breeding Station,
Plas Gogerddan, Aberystwyth, SY23 3EB, UK

ABSTRACT

Evidence is presented that the diffusion of oxygen into the bacteroid zone of legume nodules is regulated by a variable barrier. the implications of this in relation to oxygen limitation of nitrogen fixation and protection of nitrogenase during periods of environmental stress are reviewed.

INTRODUCTION

All aerobic nitrogen fixing organisms have a problem with oxygen. It is required for the production of ATP by oxidative phosphorylation but direct contact with nitrogenase leads to irreversible damage. In legume nodules the solution is to restrict oxygen supply to the bacteroid zone by means of a diffusion barrier. High rates of respiratory oxygen consumption are maintained in the bacteroid zone, with leghaemoglobin providing facilitated diffusion within the infected cells. It is now becoming generally recognized that the resistance of the diffusion barrier can be varied in a controlled manner. The implications of this diffusion control for nitrogen fixation in legumes are the subject of this review.

EVIDENCE FOR A VARIABLE BARRIER

Oxygen moves through air about 10^4 times faster than through water. Consequently air pathways which traverse the nodule offer a major route for the distribution of oxygen. In relation to these pathways, nodule structure can be divided into four zones (Witty et al., 1986):

a) the outer cortex through which oxygen can diffuse freely via a network of air pathways,

b) the inner cortex which appears to have very few air pathways,

c) the boundary between the inner cortex and the bacteroid zone which had a proliferation of interconnected pathways, and

d) the bacteroid zone which also has a substantial network of pathways.

F. O'Gara et al. (eds.), Physiological Limitations and the Genetic Improvement of Symbiotic Nitrogen Fixation, 77–85.
© *1988 by Kluwer Academic Publishers.*

Thus, the inner cortex appears to be the major impediment to diffusion. Further evidence for this has been provided by studies in which microelectrodes are gradually advanced into the bacteroid zone. For functional nodules, these show a sharp decrease in oxygen concentration across the inner cortex and very low concentrations in the bacteroid zone (Tjepkema and Yocum, 1974; Witty et al., 1987).

Indirect evidence for variability of the diffusion barrier has come from studies using intact nodulated plants, or detached nodules, in flow-through gas systems. These are summarized in Table 1.

TABLE 1. Indirect evidence for a variable diffusion barrier

Observation	Reference
1) Exposure to C_2H_2 causes a rapid decline in C_2H_4 and CO_2 production by nodules	Minchin et al., 1983b
2) This decline is reversible	Minchin et al., 1983a
3) A 4-fold increase in O_2 from 21 to 80% caused only a 36% increase in respiration, but nitrogenase was not damaged	Sheehy et al., 1983
4) Nitrogenase is damaged if O_2 is increased from 21 to 80% in less than 1 minute.	Witty et al., 1984

To understand the implications of the first three observations it is necessary to consider Fick's first law of gaseous diffusion. For nodules this can be written as:

$$F = (Oe - Oi)/R$$

where F = oxygen flux into the nodule, Oe = external oxygen concentration, Oi = internal oxygen concentration, and R = diffusion resistance.

Assuming arbitrary units of F = 10, Oe = 20%, Oi = 0% (to which it must approximate if nitrogenase is functional) and R = 2, then a decrease in F to 5 (simulating a 50% acetylene-induced decline in respiratory O_2 consumption) with Oe and R remaining constant would produce an Oi value of about 10%, resulting in nitrogenase damage. The acetylene-induced decline is reversible, indicating that nitrogenase is not damaged and Oi has stayed close to zero. This can only be achieved by a doubling of R. Similarly, an increase in Oe to 80% without a concomitant increase in respiratory oxygen consumption (i.e. F remaining constant) would, if R also remained constant, result in Oi approaching 60%. Since nitrogenase is not damaged (as evidenced by high rates of activity when Oe is returned from 80 to 21%) R must have increased substantially. The final observation concerning the speed of response of the barrier suggests that an induced change is involved.

Similar indirect evidence for a variable barrier in soyabean nodules has recently been published (Hunt et al., 1987). These authors studied changes in the production of hydrogen and carbon dioxide in response to changes in external oxygen concentration.

Direct evidence for increases in the diffusion barrier has been obtained with oxygen specific microelectrodes (Witty et al., 1987). These were advanced into the nodule until no oxygen could be detected and the external oxygen concentration was then increased to 40%. In 9 out of 21 experiments a transient increase in internal oxygen concentration of 2 to 3 minutes duration appeared on the electrode trace. This can be interpreted as an increase in oxygen flux into the inner nodule which is reduced as the barrier responds. The extra oxygen is then consumed by respiration. A transient activity increase of similar duration has also been noticed in $^{14}CO_2$ production from soyabean nodules, following a change in external oxygen concentration from 3 to 21% (Gordon et al., 1985).

Despite this evidence visual observation of the variable barrier has yet to occur and the biochemical, physiological and physical changes involved in its operation are not understood. However, it is likely that changes in the resistance of the barrier involve alterations in a water-filled pathlength through which gases must diffuse. This pathlength could be created by one or more layers of cells where all the cell walls abut and there are no air pathways. Changes in turgor pressure could lead to an expansion of these cells and an increase in the area of contact of cell walls, resulting in an increased resistance (Minchin et al., 1985b; Witty et al., 1986; Hunt et al., 1987). Alternatively, air pathways through these cell layers could occur but be occluded by water columns held between the cell walls. Variations in the length of these columns could provide the variability of the barrier. Changes in air pathway configuration have not been observed in the inner cortex, but reductions in pathways within the bacteroid zone in response to increased external oxygen concentration have been seen using dark-field microscopy (Witty et al., 1987).

DOES OXYGEN LIMIT NITROGEN FIXATION?

If the variable diffusion barrier is to allow fixation to operate at maximum capacity for a given carbohydrate supply it must maintain a very precise balance between the inflow of oxygen to the bacteroids and their respiratory requirement. Unfortunately, there is no unequivocal data which allows for an assessment as to whether the barrier accurately maintains this balance or produces oxygen limitation.

Previous experiments have demonstrated marked increases in acetylene reduction activity of detached nodules under increased external oxygen concentrations (e.g. Bergersen, 1970; Mague and Burris, 1972). However, it is now known that detachment causes an increase in the barrier resistance (Minchin et al., 1985b; Hunt et al., 1987) and oxygen-induced increases can be re-interpreted as removal of the oxygen stress imposed by detachment. Similarly, large oxygen-induced increases with intact root systems could reflect recovery from the oxygen stress resulting from acetylene-induced increases in the barrier (Witty et al., 1984). The same problem exists when measuring hydrogen production under argon, as there is also an argon-induced increase in the barrier (Witty et al., 1984; Hunt et al., 1987). Thus, the effect of changes in external oxygen concentration on in vivo nitrogenase activity cannot be measured because the available techniques cause increases in the barrier, resulting in oxygen-limiting conditions within the nodules (Witty et al., 1986).

An indirect approach can be made through the use of mathematical models using the known information on nodule structure and biochemistry, and the physics of gaseous diffusion (Sheehy et al., 1985; Sheehy and Bergersen, 1986). Such models demonstrate that, despite the presence of leghaemoglobin, there must be a small oxygen gradient across each bacteroid containing cell. Therefore, if the barrier is regulated to restrict oxygen supply so as to avoid damage to the nitrogenase at the edge of the infected cells, then the bacteroids towards the centre must be under a degree of oxygen limitation. That is, the attainable maximum fixation capacity for a given supply of carbohydrate must always be less than the theoretical maximum capacity.

This minimal level of limitation (Fig. 1A) could be increased if the diffusion barrier always restricted oxygen entry to a rate substantially less than that required for maximum enzyme activity (Fig. 1B), or the barrier maintained a balance between oxygen supply and respiratory requirement for low rates of carbohydrate supply but was unable to allow sufficient oxygen entry at higher rates (Fig. 1C).

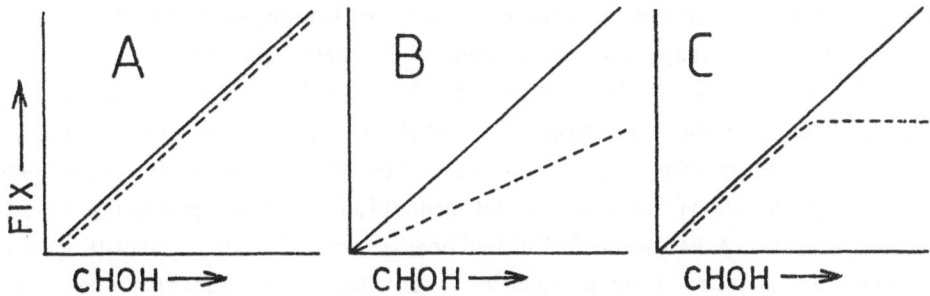

Fig. 1. Effects of different types of O_2 limitation on the response of nitrogen fixation (FIX) to increasing carbohydrate supply (CHOH). Solid lines represent theoretical maximum capacity and dashed lines represent actual rates. (A) = minimum limitation required to protect nitrogenase, (B) = limited to a fixed proportion of maximum capacity (eg 50%), (C) = minimum limitation at low rates of carbohydrate supply becoming increasingly limited at higher rates.

Experiments in which pea and soyabean have been exposed to 10, 21 or 30% oxygen for 28 days produced no significant differences in plant growth or nitrogen fixation (Minchin et al., 1985b). This might be interpreted as evidence that oxygen is not a major limiting factor. However, it is also possible that oxygen was limiting fixation by, for example, 50% of potential maximum activity at 21% oxygen and the barrier adjusted to maintain this same degree of limitation when oxygen was supplied at 10 or 30%. If such 'conservatism' of oxygen supply regulation does exist it offers an opportunity to increase rates of nitrogen fixation by 'resetting' the barrier to operate at a limitation of only 10 or 20% of potential maximum activity.

Oxygen limiting conditions would also occur if the speed of response of the barrier was too slow to take advantage of short periods of increased carbohydrate supply. Although barrier closure can be very rapid (Witty et al., 1984) opening may be much slower, at least in soyabean (Criswell et al., 1976; Hunt et al., 1987). Such transient periods of oxygen limitation may explain the build up in nodules of apparently surplus carbohydrates, especially polyhydroxybutryrate (Wong and Evans, 1971; Streeter, 1981).

OXYGEN AND ENVIRONMENTAL FACTORS

One major function of a variable diffusion barrier is to protect nitrogenase against oxygen damage during periods of environmental stress. The involvement of oxygen supply regulation in relation to water stress has been recognized for a number of years (Pankhurst and Sprent, 1975; Denison et al., 1983; Weisz et al., 1985). Recent studies with soyabean have shown a strong correlation between oxygen diffusion resistance and nodule water potential with the increase in resistance preceeding the water stress-induced reduction in photosynthesis (Durand et al., 1987). A variable diffusion barrier could also be useful during periods of reduced oxygen supply associated with waterlogging. An apparent reduction in diffusion resistance was found in nodules of cowpea grown under oxygen concentrations of 5% or less (Dakora et al., 1986). This change was accompanied by gross morphological changes similar to those observed with nodules subjected to waterlogged conditions (Minchin and Summerfield, 1976). However, comparative measurements on nodules of Neptunia plena grown in either vermiculite or unaerated nutrient solution have shown no differences in diffusion resistance, despite marked morphological variation (James et al., unpublished).

Protection of nitrogenase will also be important during stresses which reduce carbohydrate supply to nodules, such as defoliation or prolonged shoot darkening. The former appears to cause no damage to nitrogenase in white clover (Ryle et al., 1986) whilst the metabolism of bacteroids isolated from soyabean nodules is little affected during four days of darkness (Sarath et al., 1986; Carroll et al., 1987). Both these observations can be explained by an increase in diffusion resistance. Indeed, increases in the barrier have been reported for plants subjected to defoliation (Minchin et al., 1985b; Hartwig et al., 1987) and 24 h shoot darkening (Minchin et al., 1985a; Carroll et al., 1987). Minchin et al. (1985b) also noted different rates of resistance increase in response to total shoot removal between white clover (very rapid), soyabean (intermediate) and pea (no response for first hour). This may be related to adaptation to defoliation in these species. Carbohydrate supply and utilization is also reduced by low temperatures (Layzell et al., 1984), but nitrogenase is not damaged and this stress must be accompanied by an increase in the diffusion barrier (Witty et al., 1986). Again, Minchin et al. (1985b) noted different rates of response to a 10°C temperature decrease between white clover and pea.

Changes in resistance have also been implicated in recent studies on the effect on nitrate on nitrogen fixation of white clover and soyabean

(Minchin et al., 1986; Carroll et al., 1987). A decrease in oxygen supply could be the major factor responsible for the reduction in nitrogenase activity during the early stages of nitrate stress. This hypothesis is supported by reports of a significant nitrate effect on nitrogen fixation by nodules, but a lack of effect on the subsequently isolated bacteroids (Houwaard, 1980; Schuller et al., 1986), and recent work indicating that nitrate enters the nodule cortex but not the bacteroid zone (Sprent et al., 1987).

CONCLUSION

The resistance of the oxygen diffusion barrier of legume nodules can vary in response to a wide range of stimuli (Table 2) and the speed and extent of variation depends upon both Rhizobium and host plant genotype (Minchin et al., unpublished). The accuracy with which oxygen flux can be balanced against carbohydrate supply will determine whether oxygen will (a) limit nitrogen fixation, (b) exactly balance respiratory requirements or (c) damage nitrogenase. An ability to understand and manipulate the functioning of the diffusion barrier may lead to stable, or even increased, nitrogen fixation by legume crops.

TABLE 2. Factors affecting the nodule diffusion barrier

External oxygen concentration
Cessation of ammonium production (due to exposure to acetylene or argon)
Disturbance or detachment of nodules
Water stress
Waterlogging
Defoliation
Prolonged darkness
Low temperature
Nitrate

REFERENCES

Bergersen, F.J. 1970. The quantitative relationship between nitrogen fixation and the acetylene-reduction assay. Aust. J. Biol. Sci., 23, 1015-1025.
Carroll, B.J., McNeil, D.L. and Gresshoff, P.M. 1987. Oxygen supply to nodules limits nitrogen fixation (acetylene reduction) in nitrate- and dark-inhibited soybean (Glycine max (L.) Merr.) plants. Planta (in press).

84

Criswell, J.G., Havelka, U.D., Quebedeaux, B. and Hardy, R.W.F. 1976. Adaptation of nitrogen fixation by intact soybean nodules to altered rhizosphere pO_2. Plant Physiol., 58, 622-625.

Dakora, F.D., Atkins, C.A. and Pate, J.S. 1986. Oxygen sensitivity of nitrogenase and N_2 fixation in cowpea nodules developing in sub-ambient O_2 levels. In 'The Eighth Australian Nitrogen Fixation Conference' (Ed. W. Wallace and S.E. Smith). (AIAS, Canberra), pp. 11-12.

Denison, R.F., Weisz, P.R. and Sinclair, T.R. 1983. Analysis of acetylene reduction rates of soybean nodules at low acetylene concentrations. Plant Physiol., 73, 648-651.

Durand, J-L., Sheehy, J.E. and Minchin, F.R. 1987. Nitrogenase activity, photosynthesis and nodule water potential in soyabean plants experiencing water deprivation. J. Exp. Bot., 38, 311-321.

Gordon, A.J., Ryle, G.J.A., Mitchell, D.F. and Powell, C.E. 1985. The flux of ^{14}C-labelled photosynthate through soyabean root nodules during N_2 fixation. J. Exp. Bot., 36, 756-769.

Hartwig, U., Boller, B. and Nosberger, J. 1987. Oxygen supply limits nitrogenase activity of clover nodules after defoliation. Ann. Bot., 59, 285-291.

Houwaard, F. 1980. Influence of ammonium and nitrate nitrogen on nitrogenase activity of pea plants as affected by light intensity and sugar addition. Plant Soil., 54, 271-282.

Hunt, S., King, B.J., Canvin, D.T. and Layzell, D.B. 1987. Steady and nonsteady state gas exchange characteristics of soybean nodules in relation to the oxygen diffusion barrier. Plant Physiol., 84, 164-172.

Layzell, D.B., Nochman, P. and Canrin, D.T. 1984. Low root temperatures and nitrogenase activity in soybean. Can. J. Bot., 62, 965-971.

Mague, T.H. and Burris, R.H. 1972. Reduction of acetylene and nitrogen by field-grown soybeans. New Phytol., 71, 275-286.

Minchin, F.R., Minguez, M.I., Sheehy, J.E., Witty, J.F. and Skot, L. 1986. Relationships between nitrate and oxygen supply in symbiotic nitrogen fixation by white clover. J. Exp. Bot., 37, 1103-1113.

Minchin, F.R., Sheehy, J.E., Minguez, M.I. and Witty, J.F. 1985a. Characterisation of the resistance to oxygen diffusion in legume nodules. Ann. Bot., 55, 53-60.

Minchin, F.R., Sheehy, J.E. and Witty, J.F. 1985b. Factors limiting N_2 fixation by the Legume-Rhizobium symbiosis. In 'Nitrogen Fixation Research Progress' (Ed. H.J. Evans, P.J. Bottomley and W.E. Newton). (Martinus Nijhoff, Dordrecht), pp. 285-291.

Minchin, F.R. and Summerfield, R.J. 1976. Symbiotic nitrogen fixation and vegetative growth of cowpea (Vigna unguiculata (L.) Walp.) in waterlogged conditions. Plant Soil, 45, 113-127.

Minchin, F.R., Witty, J.F. and Sheehy, J.E. 1983a. A new technique for the measurement of respiratory costs of symbiotic nitrogen fixation. In 'Temperate Legumes: Physiology, Genetics and Nodulation' (Ed. D.G. Jones and D.R. Davies). (Pitman, London). pp. 201-217.

Minchin, F.R., Witty, J.F., Sheehy, J.E. and Muller, M. 1983b. A major error in the acetylene reduction assay: decreases in nodular nitrogenase activity under assay conditions. J. Exp. Bot., 34, 641-649.

Pankhurst, C.E. and Sprent, J.I. 1975. Effects of water stress on the respiratory and nitrogen-fixing activity of soybean root nodules. J. Exp. Bot., 26, 287-304.

Ryle, G.J.A., Powell, C.E. and Gordon, A.J. 1986. Defoliation in white clover: nodule metabolism, nodule growth and maintenance, and nitrogenase functioning during growth and regrowth. Ann. Bot., 57, 263-271.

Sarath, G., Pfeiffer, N.E., Sodh, C.S. and Wagner, F.W. 1986. Bacteroids are stable during dark-induced senescence of soybean root nodules. Plant Physiol., 82, 346-350.

Schuller, K.A., Day, D.A., Gibson, A.H. and Gresshoff, P.M. 1986. Enzymes of ammonium assimilation and ureide biosynthesis in soybean nodules: effect of nitrate. Plant Physiol., 80, 646-650.

Sheehy, J.E. and Bergersen, F.J. 1986. A simulation study of the functional requirements and distribution of leghaemoglobin in relation to biological nitrogen fixation in legume root nodules. Ann. Bot., 58, 121-136.

Sheehy, J.E., Minchin, F.R. and Witty, J.F. 1983. Biological control of the resistance to oxygen flux in nodules. Ann. Bot., 52, 565-571.

Sheehy, J.E., Minchin, F.R. and Witty, J.F. 1985. Control of nitrogen fixation in a legume nodule: an analysis of the role of oxygen diffusion in relation to nodule structure. Ann. Bot., 55, 549-562.

Sprent, J.I., Giannakis, C. and Wallace, W. 1987. Transport of nitrate and calcium into legume root nodules. J. Exp. Bot., 38, 1121-1128.

Streeter, J.G. 1981. Seasonal distribution of carbohydrates in nodules and stem exudate from field-grown soya bean plants. Ann. Bot., 48, 441-450.

Tjepkema, J.D. and Yocum, C.S. 1974. Measurement of oxygen partial pressure within soybean nodules by oxygen microelectrodes. Planta, 119, 351-360.

Weisz, P.R., Denison, R.F. and Sinclair, T.R. 1985. Response to drought stress of nitrogen fixation (acetylene reduction) rates by field-grown soybeans. Plant Physiol., 78, 525-530.

Witty, J.F., Minchin, F.R., Sheehy, J.E. and Minguez, M.I. 1984. Acetylene-induced changes in the oxygen diffusion resistance and nitrogenase activity of legume root nodules. Ann. Bot., 53, 13-20.

Witty, J.F., Minchin, F.R., Skot, L. and Sheehy, J.E. 1986. Nitrogen fixation and oxygen in legume root nodules. In 'Oxford Surveys of Plant Molecular & Cell Biology, Vol. 3' (Ed. B.J. Miflin). (Oxford Press, Oxford). pp. 275-314.

Witty, J.F., Skot, L. and Revsbech, N.P. 1987. Direct evidence for changes in the resistance of legume root nodules to O_2 diffusion. J. Exp. Bot., 38, 1129-1140.

Wong, P.P. and Evans, H.J. 1971. Poly-β-hydroxybutyrate utilization by soybean (Glycine max Merr.) nodules and assessment of its role in maintenance of nitrogenase activity. Plant Physiol., 47, 750-755.

FACTORS CONTROLLING LEGUME NITROGEN FIXATION AND THEIR MEASUREMENT IN THE FIELD

J.E. Sheehy[*] and A. McNeill[**]

[*]The Institute for Grassland and Animal Production,
Hurley, Maidenhead, Berks., UK. and
[**]The Department of Soil Science,
University of Reading,
Reading, Berks., UK.

ABSTRACT
   Attention is drawn to our lack of understanding of the factors limiting nitrogen fixation in the field. A flow through apparatus for measuring nitrogenase activity in field crops is described. A marked seasonal variation in nitrogenase activity in sainfoin was observed which was, in part, the result of the effect of defoliation . The acetylene based estimate of the nitrogen fixed annually by sainfoin was 146 kg ha^{-1} less than the nitrogen accumulated in the shoots. There were no treatment differences between the rates of growth of either sainfoin or lucerne when grown with or without fertiliser nitrogen in the field.

INTRODUCTION

   Genetic and environmental factors determine the charact-

eristics which shape nodule function. Recent developments in

molecular biology have stimulated an enormous increase in re-

search concerned with Rhizobium genetics. One method of using

this powerful tool would be to investigate the degree to which

genetic factors control those aspects of nodule function which

limit symbiotic nitrogen fixation in agriculture. Although it

is well known that soil and weather conditions affect

symbiotic nitrogen fixation, the precise manner in which those

conditions affect the efficiency, capacity and rate of that

process remain obscure. Thus, the factors limiting symbiotic

nitrogen fixation in practice remain a matter for speculation.

Indeed, a major objective of this paper is to draw attention

F. O'Gara et al. (eds.), Physiological Limitations and the Genetic Improvement of Symbiotic Nitrogen Fixation, 87–96.
© 1988 by Kluwer Academic Publishers.

to our lack of knowledge concerning nodule function in the field. In part, this deficiency in our understanding stems from the absence of appropriate techniques for use in the field.

Work in the controlled environment (see Minchin et al, this volume) has demonstrated the importance of nodule gaseous diffusion resistance in regulating oxygen diffusion and carbohydrate utilisation in response to changes in environmental conditions. There are of course difficulties in extrapolating results results from controlled environments to field situations. Ritchi (1981) observed that the relationship between stomal resistance and water potential for field grown cotten was markedly different from that observed in plants grown in a controlled environment. A similar divergence in the relationship between leaf extension and temperature was observed by Watts (1974). It is not known whether there are such differences in nodule characteristics between field and controlled environment grown legumes.

The overall objective of this paper is to examine some aspects of nodule formation in the field and to describe the use of a field apparatus for investigating the seasonal variation in nitrogenase activity. It should be noted that (a) the apparatus was used in sainfoin to avoid problems relating to the acetylene error described by Minchin et al. (1983) and (b) the sainfoin-strain combination used does not evolve hydrogen.

MATERIALS AND METHODS

Lucerne and sainfoin crops were both established during summer. The seed of lucerne variety Vertus was drilled in rows 100 mm apart at a rate of 20 kg ha^{-1}. A Hampshire Giant sainfoin was sown at a rate of 300 kg ha^{-1} of milled ssed. The lucerne seed was inoculated with Nodulaid "A"; there was no requirement for sainfoin inoculant. Two nitrogen treatments were imposed. Half of the plots in each crop were given nitrate fertiliser at a rate of 750 kg N ha^{-1} (300 kg in spring and 150 kg after each cut). The other half did not receive nitrate. P and K fertilisers were applied to all plots at an annual rate of 70 kg P ha^{-1} and 430 kg K ha^{-1} , fifty percent being applied at the beginning of the spring growth period.

The apparatus for investigating acetylene reduction was only used in sainfoin. It consisted of: (a) a perspex shoot enclosure, (b) an air-conditioning and mixing system, (c) a rim (0.5 m x 0.5 m x 0.18m) inserted into the soil, (d) a grid for gas delivery situated at the bottom of the rim and (e) various pumps and flow meters for delivering air to and removing air from the apparatus.

Four rims and associated grids were inserted into the zero-nitrate sainfoin treatment. The apparatus was used as described by Stiles (1977) to measure shoot photosynthesis and respiration by passing air into the perspex enclosures so that a slight positive pressure was created which excluded "soil" CO_2. Acetylene reduction was measured by pumping an pumping an acetylene air mixture in through the grid and sucking it from the perspex enclosure.

The continuity equation was used to estimate the rates of acetylene reduction, respiration and photosynthesis. The enclosed soil and the enclosed shoots were considered as two separate, but connected enclosures. Thus the output from the soil was the input from the shoot encosure. Rates of acetylene reduction were made in the steady state (no acetylwne decline with sainfoin), as were rates of "soil" respiration. To achieve rapid mixing air was passed into the soil grid at 20 1 min^{-1}. Approximately 95% of the acetylene entering the grid was present at the output from the shoot enclosure; the precise fate of the 5% remains to be determined. It was assumed that the reduction of four moles of acetylene was equivalent to the reduction of one mole dinitrogen.

Individual plants were grown in a glsshouse. Inoculated sainfoin seed was grown on a tray and kept moist with demin-eralised water. Eight days after sowing the seedlings were transplanted to 9.5 cm diameter pots and re-inoculated. The seedlings were given an N-free nutrient solution and tap water (containing 8 mg N 1^{-1}) on alternate days until the fourth leaf stage. Therafter, nutrient solutions containing different quantities of nitrate were applied daily (zero, 22 and 220 mg N 1^{-1}). Five plants from each treatment were harvested thir-teen weeks after germination and the nitrogen contents were determined using Kjeldahl analysis.

RESULTS AND DISCUSSION

The dry composition of sainfoin plants receiving different amounts of nitrate nitrogen are shown in Table 1. The

Table 1   Sainfoin plants utilising 221, 22 mg ℓ^{-1} of nitrate-
nitrogen and biological nitrogen fixation, harvested
13 weeks after germination.

Concentration N (mg ℓ^{-1})	221	22	0	Mean S.E.
Shoot d. wt (g)	4.416	1.987	1.151	± 0.90
Root d. wt (g)	1.484	1.040	0.630	± 0.39
Nodule d. wt (g)	0.021	0.189	0.212	± 0.06
Total Plant d wt	5.92	3.36	2.00	± 0.97
Shoot:root	3.18	1.73	1.36	± 0.6
% Flowering	100	100	zero	−
Leaf Area (cm^{-2} $plant^{-1}$)	391.5	224.5	134.7	± 43.65

smallest dry weights and leaf area were recorded in plants not
receiving nitrate; acetylene reduction rates were greatest in
this treatment (McNeill, 1984). The addition of 220 mg N 1^{-1}
effectively reduced acetylene reduction to zero and almost
completely suppressed nodule formation; those nodules formed
lacked leghemoglobin and were virtually ineffective (McNeill,
1984). The addition of 22 mg N 1^{-1} had no significant effect
on acetylene reduction or the dry weight of nodules. However,
it did have a marked effect on flowering. Measurements of respiration made on individual plants suggest that nodule respiration accounts for approximately 17% of the carbohydrate assimilated daily (Table 2).

Table 2   Daily assimilate utilization expressed as a %, in
individual sainfoin plants of mean weight 6.4 mg and
leaf area (including petiole) of 299 cm^2

Shoot respiration	24
Root respiration	17
Nodule respiration	17
Unrespired	42

It would seem reasonable to expect the costs of symbiotic nitrogen fixation to reduce the growth rates of legumes fixing
dinitrogen relative to those reducing nitrate in their leaves
using light energy (Sheehy et al., 1984). Such a reduction was
not observed in the field crops of sainfoin or lucerne.

There were no significant differences between the dry
matter yields of sainfoin growing in the field with or without
the addition of nitrate (Table 3). A similar result was

Table 3   Above-ground dry matter at the end of each growth period
           for an unirrigated sainfoin crop grown with or without
           fertiliser nitrogen.

	Above ground d.m[1] (kg ha^{-1})	
	+N	-N
1st growth period	7203	8203
2nd growth period	5695	5216
3rd growth period	2922	2753

No significant ($P < 0.05$) differences between +N and -N treatments

obtained for lucerne and the maximum rates of nitrogen accumulated in the shoots were not significantly different (Table 4).

Table 4   Mean nitrogen increment rates in lucerne during linear
           phase (kg ha^{-1} day^{-1})

GROWTH PERIOD	+N	-N
First (weeks 5-8)	4	4
Second (weeks 13-17)	5	4
Third (weeks 21-23)	6	6

No significant differences ($P < 0.05$)

The results indicate that field grown legumes do not appear to
suffer any yeild or quality loss as a consequence of symbiotic
nitrogen fixation. Measurements of the seasonal pattern of dinitrogen fixation were made in the field grown sainfoin.

During the first growth period the rate increased from approximately 0.3 kg N ha^{-1} day^{-1} in March to 4.2 kg N ha^{-1} day^{-1} in June. Following defoliation nitrogenase activity declined rapidly; three hours after defoliation it had declined by 71% and was zero forty hours later. The nitrogenase activity did not show any recovery for a further two weeks, thereafter it recovered gradually to reach 1.1 kg N ha^{-1} day^{-1}. Following a second defoliation nitrogenase activity recovered very slowly over a period of approximately ten weeks to 1.2 kg N ha^{-1} day^{-1}.

During the first growth period the nitrogen accumulated in the shoots was 58 kg N ha^{-1} less than the acetylene based estimate. In the second growth period the shoots accumulated 94 kg N ha^{-1} more than was estimated using the acetylene technique and in the third growth period the technique underestimated by 110 kg N ha^{-1}. The total nitrogen accumulated by the sainfoin shoots was 416 kg N ha^{-1} year^{-1} whereas the amount of nitrogen estimated to be fixed was 270 kg N ha^{-1} year^{-1}. The difference between observed nitrogen accumulation and the estimated rate of nitrogen was due to a marked shortfall during the summer and autumn growth periods.

If the field acetylene technique can be relied upon, than it would appear that during the second and third growth periods regrowth of the shoots depends on either stored nitrogen or soil nitrogen. If on the other hand the quantitative estimates of nitrogen fixation are contentious it is interesting to note that a similar change in the seasonal pattern

of nitrogen fixation was reported by Newbould and Haysted
(1978) for white clover growing in a brown earth soil. It is
difficult to compare absolute rates as no estimate of the
effect of the acetylene induced decline in nitrogenase activity on the white clover results can be made. Nevertheless,
the pattern suggests that the rate of activity is greater in
spring than in summer or autumn.

The explanation for the decline can not be found in a
shortfall in photosynthesis (Table 5); The average daily rate

Table 5   Mean total daily photosynthesis and nodule respiration as
a % of photosynthesis (McNeill 1984)

Growth Period	Photosynthesis $(g\ CO_2\ m^{-2}\ day^{-1})$	Nodule respiration[+] (%)
27/3 - 22/5	36.0	23.7
3/7 - 10/7	39.8	5.7
7/8 - 11/9	21.8	2.9

[+] (4.93 mole CO_2/mole C_2H_4)

of photosynthesis was greater in summer than in spring. Furthermore the estimated fraction of photosynthate used for
nitrogen fixation decreased from 23.7% in spring to 2.9% in
autumn. The nitrogen contents of the soil under the lucerne
treatments were determined and as shown in Table 6 there are
marked differences. Nitrogen levels in the soils not receiving
nitrate would severely limit crop growth; annual dry matter
yields of approximately 50 kg ha^{-1} might be expected (Cowling,
1982).

It is clear that the seasonal pattern of symbiotic nitrogen fixation in the field requires further investigation.
Legume plants possessing a controlled nodule diffusion resist-

Table 6   Means of soil nitrogen results in lucerne (kg ha^{-1})

Growth period	Profile	+N Total N	-N Total N
First	0-15	94.6 ± 14.0	5.5 ± 0.4
	15-30	18.1 ± 2.7	3.8 ± 0.2
Second	0-15	86.2 ± 14.1	8.9 ± 0.4
	15-30	44.7 ± 13.6	8.1 ± 1.4
Third	0-15	76.6 ± 11.4	8.4 ± 0.5
	15-30	11.1 ± 2.1	5.5 ± 1.0

ance may well exhibit a different defoliation response to

sainfoin. To investigate nitrogenase activity in these other

legume crops, field technique requires considerable

refinement.

REFERENCES

Cowling, D.W. 1982. Biological nitrogen fixation and grassland
    production in the United Kingdom. Phil. Trans. R. Soc.
    Lond. B296, 397-404.
McNeill, A. 1984. Environmental effects on the physiology of
    sainfoin (Onobrychis viciifolia), with particular refer-
    ence to nitrogenase activity. Ph. D. Thesis, University
    of Reading.
Minchin, F.R., Witty, J.F., Sheehy, J.E. and Muller, M. 1983.
    A major error in the acetylene reduction assay: Decrease
    in nodular nitrogenase activity under assay conditions.
    J. Exp. Bot. 34, 641-649.
Newbould, P. and Haysted, A. 1978. Trifolium repens (white
    clover): its role, establishment and maintenance in hill
    pastures. Hill Farming Research Organisation Report
    Number 7, 49-68.
Ritchie, J.T. 1981. Water dynamics in the soil-plant-atmos-
    phere system. Plant and Soil, 58, 81-96.
Sheehy, J.E., Minchin, F.R. and McNeill, A. 1984. Physiologi-
    cal principles governing the growth and development of
    lucerne, sainfoin and white clover. Occasioal Symposium
    16, British Grassland Society, pp. 112-126.
Stiles, W. 1977. Enclosure method for measuring photosynthesis,
    respiration and transpiration of crops in the field.
    Grassland Research Institute Technical Report No. 18.

Watts, W.R. 1974. Leaf extension in zea Maize. III. Field
    measurements of response to temperature and leaf water
    potential. J. Exp. Bot. 25, 1088-1096.
Witty, J.F., Minchin, F.R. and Sheehy, J.E. 1983. Carbon costs
    of nitrogenase activity in legume nodules determined
    using acetylene and óxygen. J. Exp. Bot. 34, 951-963.

INHIBITION OF SYMBIOTIC NITROGEN FIXATION BY NITRATE

M.O. HECKMANN, J.J.DREVON,
Institut National de la Recherche Agronomique
Laboratoire de Recherche sur les Symbiotes des Racines
34000 Montpellier , France

ABSTRACT
Symbiotic nitrogen fixation is inhibited in the presence of nitrate. Various symbiotic processes are affected including nitrogenase activity which decreases just after addition of nitrate to the medium. Experimental results suggest that this inhibition is not due to reduction of nitrate in the shoots and roots but to nitrate ion metabolism in root nodules where bacterial reduction of nitrate does not appear to play a major role. Considerable nitrate metabolism was found to occur in the plant compartment of the nodules. This could cause inhibition by nodular accumulation of nitrite produced by cytosolic nitrate reductase activity and by competition with bacteroids for the reduction energy.

INTRODUCTION

Legumes in symbiosis with bacteria of the genera *Rhizobium* and *Bradyrhizobium* have two sources of nitrogen nutrition : fixation of atmospheric molecular nitrogen and assimilation of soil nitrogen , found mostly in the form of nitrate.

When both nitrogen sources are available, the contribution of N_2 fixation to the nitrogen nutrition of the plant varies with growth conditions , the plant genotype (Duc, 1980) and the combined nitrogen level in the soil (Bello et al., 1980). In soybean, for a yield of 20-30 qx/ha, the percentage of fixed nitrogen varies from 25 % in highly fertile soils to 80 % in infertile soils (La Rue and Patterson, 1981). In Alfalfa, 40 to 70 % of total nitrogen results from symbiosis (Heichel et al., 1981). Vicia faba minor can receive 87 % of its total nitrogen from N_2 fixation (Richard and Soper, 1978).

Legumes apparently cannot fix sufficient nitrogen for maximum growth response (Harper, 1974) which demonstrates the need for combined nitrogen (Harper, 1975). Moreover, the study of the effects of nitrogen supply on productivity in nodulating (N_2 dependent) and non-nodulating (NO_3^- dependent) legumes suggests that NO_3^- assimilation produces a higher yield than N_2 fixation (Salsac et al., 1984), although in soybean, the highest yield was attained by the combination of both nitrogen sources (Harper, 1974).

However, NO_3^- assimilation by legumes can inhibit N_2 fixation; it stops nodulation and decreases nitrogenase activity (Gibson, 1976). This process, which is not yet completely understood, prevents maximum exploitation of nitrogen fixation in well fertilized soils. This is the case in most temperate regions.

F. O'Gara et al. (eds.), Physiological Limitations and the Genetic Improvement of Symbiotic Nitrogen Fixation, 97–106.
© 1988 by Kluwer Academic Publishers.

The scope of this paper will be limited to the investigation of short-term reversible inhibition of nitrogenase activity by nitrate . First we consider the role of nitrate metabolism in the aerial parts of legumes and in roots, root nodules, plant cells and bacteria, with the emphasis on the contribution of nodule cytosol nitrate reductase. Second, we discuss the mechanisms of nitrogenase activity inhibition by nitrate metabolism in the nodules.

NITRATE METABOLISM

Reduction of nitrate in shoots and roots

Nitrate absorbed by legumes roots can be assimilated via an enzymatic pathway common to the shoots and the roots (Beevers and Hageman, 1980). The reduction of NO_3^- to NO_2^-, catalyzed by the nitrate reductase (NR) is generally considered to be the limiting factor in NO_3^- assimilation (Beevers and Hageman, 1980). In higher plants, this enzyme, which is induced by NO_3^-, is generally specific for NADH (Beevers and Hageman, 1980). However, in soybean leaves, there is also a bispecific NAD(P)H-NR which can receive electrons from NADPH (Campbell, 1976).

During the cycle of field-grown plants, NO_3^- assimilation complements N_2 fixation to meet the nitrogen requirements of the symbiosis (Obaton et al., 1982) i.e combined nitrogen is essential in the early life of most legumes, after the nitrogen content of the seed has been exhausted and before nitrogen fixation is sufficiently active. The decrease in nitrogenase activity during pod filling is generally interpreted as a diversion of photosynthates to the pods at the expense of the nodules (Lawn and Brun, 1974). It thus appears that the relay of N_2 fixation by NO_3^- assimilation is facilited by application of nitrogen fertilizer during later growth stages. But this succession cannot be generalized, since the position of the activity peak varies with the form and the level of nitrogen in the soil (Buttery, 1986), the host plant (Mague and Burris, 1972), and the *Rhizobium* strain (Rennie and Kemp, 1983).

Since nitrate reduction consumes large amounts of energy (Salsac et al., 1984) and is localized mostly in the aerial parts (Pate, 1973), it was suggested at an early stage that the consumption of reducing equivalents by NO_3^- metabolism in the leaves of a nodulated legume would deprive the nodules of their carbohydrate supply and cause nitrogenase activity to decrease (Oghoghorie and Pate, 1971).

But this hypothesis is presently under review since an increase in CO_2 does not prevent the inhibition of nitrogenase although it increases the supply of photosynthates to bacteroids, and application of NO_3^- on the leaves increases NR activity without inhibiting nitrogenase activity

(Chen and Phillips, 1977). Moreover, nitrate really must be in contact with nodules to induce nitrogenase inhibition, since deep application of combined nitrogen allowes greater root nitrate uptake than uniform incorporation, without concomitant inhibitory effects on symbiotic activity (Harper and Cooper, 1971). From these observations, it appears doubtful that shoot and root reduction of NO_3^- could be the cause of the nitrogenase inhibition. It is more likely due to NO_3^- metabolism in the nodules, where the two processes, N_2 fixation and NO_3^- assimilation, take place.

Nodular reduction of nitrate

Inside the nodules, nitrate assimilation can take place in two different compartments, i.e plant cell cytoplasm (cytosol) and *Rhizobia*. Nitrate metabolism in soybean nodules has been characterized by the nitrate reductase activity of both participants in the symbiosis : the NR activities of the bacteroids and nodular cytosol were found to be comparable and significantly higher than those of the roots (Heckmann and Drevon, 1987 ; Table 1). Nodular NR activity has also been shown to be higher than that of roots by Hunter (1983) and Becana et al. (1985) although these authors attributed it mostly to bacteroidal NR activity.

TABLE 1   Nitrate and nitrite reductases activities in the 40 day old soybean leaves, roots and cytosol and bacteroids. Results are expressed in μmol NO_2^-produced (NR) or consummed (NiR).$h^{-1}.g^{-1}$fresh weight.

	NR activity	NiR activity	References
	(μmol NO2- . h-1 . g-1 FW)		
Leaves (NADH-NR)	9.45		Conejero et al., 1984 .
Roots	0.25 ± 0.02	1.00 ± 0.05	Heckmann and Drevon, 1987.
Cytosol	2.40 ± 0.06	6.00 ± 0.54	
Bacteroids	3.64 ± 0.09	2.08 ± 0.10	

A nitrite concentration of about 0.5 mM has been measured in nodules with high NR activity (Streeter, 1982; Becana et al., 1985). It increases to about 1 mM during pod filling and maturation, although on the same plants, root NO_2^- content remains at a very low level (Heckmann and Drevon, 1987).

In vivo, nitrite content in soybean root nodules decreases with increasing partial pressures of

O_2 in the incubation chamber (Heckmann and Drevon, 1987). Thus, nitrite reductase (NiR) activity appears to be somewhat inhibited *in vivo* by partial anaerobiosis inside the nodules. In this process, NO_2^- reduction appears to be linked to the oxidative pentose phosphate pathway

(OPPP) which produces NADPH from glucose-6-phosphate (G6P). At low O_2 concentrations, such as in the nodules (Tjepkema and Yocum, 1974), the "Pasteur" effect probably causes a decrease in the G6P concentration and a reduction of the entry of this substrate into the leucoplastids, which then decreases the production of reduction energy from the OPPP and consequently causes the inhibition of NiR activity (Heckmann and Drevon, 1987).

This accumulation of nitrite produced by the NR activity of the cytosol and the bacteroids, could be the cause of the inhibition of N_2 fixation, since NO_2^- is an inhibitor of nitrogenase (Trinchant and Rigaud, 1980) and also of leghemoglobin (Rigaud and Puppo, 1977).

Using deficient NR strains (NR^- mutants) of *Rhizobium*, the above mentioned hypothesis has been confirmed with *in vitro* bacteroids (Stephens and Neyra, 1983) and free living nitrogen fixing *Bradyrhizobia* (Keister and Evans, 1976). Nitrogenase activity is inhibited by NO_3^- in strains with NR activity, but the inhibition is weaker in NR^- mutants (Stephens and Neyra, 1983). However, this is not the case *in situ* . In the soybean-*Bradyrhizobium japonicum* symbiosis, the suppression of the bacteroidal NR activity (NR^- mutants) decreased NO_2^- accumulation by

43 % and the nitrogenase inhibition only by 17 % (Table 2). According to Streeter (1985), removing bacterial NR (NR^- mutants) consistently decreases the nodules NO_2^- concentration without significantly decreasing nitrogenase inhibition. NO_3^- reduction in bacteroids could be involved in the nitrogenase inhibition process, but the major inhibitory effect appears to be due to cytosolic nitrate metabolism.

TABLE 2   Effect of nitrate (3 mM) on the NO_2^- concentration and the inhibition of nitrogenase activity of nodules induced by the strains PJ17 nal (NR^+) and II169 (NR^- mutant) of *Bradyrhizobium japonicum* . Activities are expressed in percentage of that of control.

	NO2- concentration (mM)	Nitrogenase inhibition (% control)
PJ17 nal	1,15	61,5
II169	0,64	51

MECHANISMS OF NITROGENASE INHIBITION BY NODULAR NO_3^- METABOLISM

Nitrite which accumulates in the nodules could cause the inhibition of nitrogenase activity by decreasing the supply of O_2 to the bacteroids, thereby affecting nitrogen fixation (Trinchant and Rigaud, 1980) since the O_2 concentration of the air appears to be a limiting factor of nodule nitrogenase activity (Ralston and Imsande, 1982; Drevon et al., 1988).

However, it is doubtful that nitrite accumulation inside the nodules is solely responsable for the inhibition of nitrogen fixation in the presence of nitrate : nitrogenase activity is less stimulated by suboptimal partial pressures of O_2(pO_2) in the presence of nitrate than in untreated plants (Fig. 1), although NO_2^- accumulation is decreased (Heckmann and Drevon, 1987).

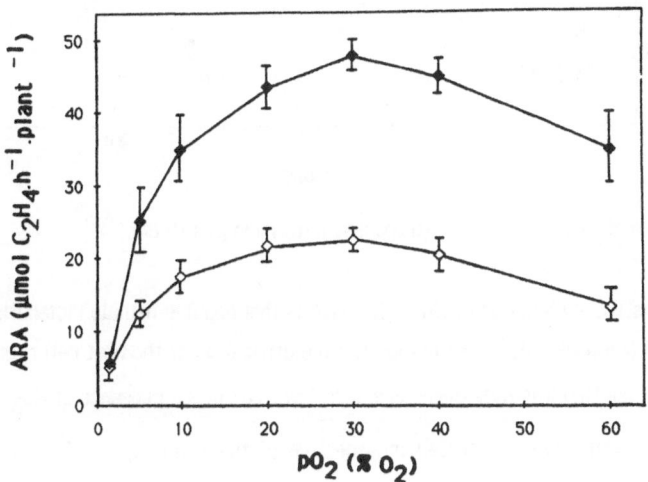

Fig. 1 Influence of the pO_2 on the *in situ* nodule acetylene reducing activity (ARA).
The ARA was estimated on 40 day old soybeans (*Glycine max* cv Hodgson- *B. japonicum* strain PJ17nal) grown in presence ($\diamond$) or in absence ($\blacklozenge$) of NO_3^- (3 mM).

According to previous studies, nitrite appears to inhibit nitrogenase and leghemoglobin (Trinchant and Rigaud, 1980) but the removal of nitrate from the preincubation mixture resulted in the complete recovery of C_2H_2 reduction, indicating a completely reversible binding between nitrogenase and NO_2^- (Fig. 2).

102

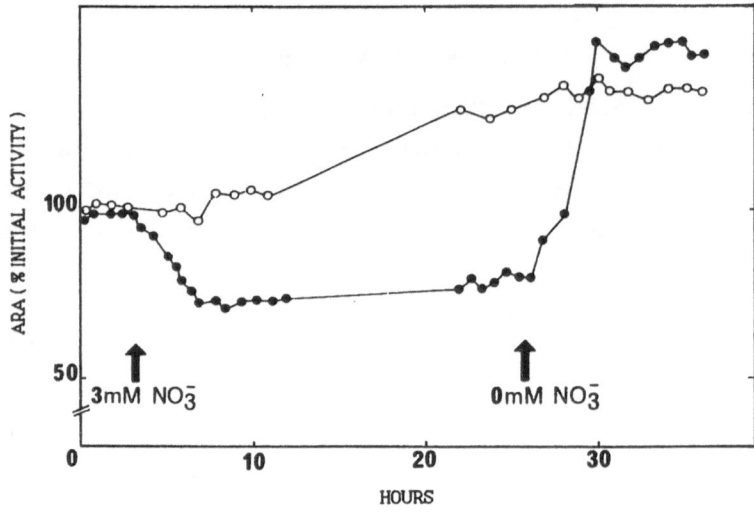

Fig. 2 Kinetics of soybean nodules nitrogenase inhibition by nitrate.

Thus an alternative explanation of this inhibition is that nodular nitrate metabolism competes with mitochondrial and bacteroidal respiration for the utilization of the host cell reducing energy pool. Arguments supporting this hypothesis are as follow : (a) in the presence of NO_3^-, the energy charge of soybean nodules was depressed irrespective of the level of pO_2 (Heckmann et al., unpublished results), (b) ammonium inhibits N_2 fixation, though less drastically than nitrate at a similar concentration (Kimou et al., 1985), (c) inhibition by NH_4^+ is decreased in the presence of methionine sulfoximine, which is an inhibitor of glutamine synthetase (Houwaard, 1979), (d) nitrogenase-linked respiration is less stimulated by external oxygen in the presence of nitrate (Minchin et al., 1986).

Malate is considered to be a major source of reducing energy for bacteroids (Ronson and Primrose, 1979) although these organelles can also use glucose (Trinchant and Rigaud, 1981). Thus, the addition of malate increased the ARA of soybeans grown without nitrate and it almost completely eliminated the inhibition by NO_3^- (Table 3). Consequently, nitrate appears to inhibit nitrogenase activity by limiting the availability of energetic substrates for the bacteroids, which could be the consequence of competition for reducing energy between NO_3^- assimilation and malate synthesis.

TABLE 3   Effect of malate and CO_2 on the nodule nitrogenase (C_2H_2 reduction) inhibition by nitrate. Activities are expressed in percentage of that of control.

TREATMENT	ARA (% CONTROL)
Control	100.0 + 1.2
15 mM L-malate	133.4 + 6.7
4600 ppm CO2	154.2 + 6.9
3 mM NO3-	56.2 + 4.1
3mM NO3- + 15mM L- malate	91.5 + 11.1
3mM NO3- + 4600ppm CO2	64.8 + 5.6

This interpretation is supported by observations of CO_2 effect. Addition of CO_2 did not stimulate the ARA of plants grown with NO_3^-, although it increased the ARA of untreated plants (Table 3).

Addition of CO_2 appears to stimulate phosphoenolpyruvate (PEP) carboxylase activity increasing the flow of malate to bacteroids and consequently nitrogenase activity. The presence of nitrate could limit this effect in two ways : (a) oxaloacetate from CO_2 fixation could be metabolized into aspartate for reduced nitrogen export rather than being reduced into malate, (b) nitrate reductase activity could compete with malate dehydrogenase for NADH (Fig. 3).

CONCLUSION

When legumes are cultivated in natural field conditions, symbiotic fixation of nitrogen by inoculated plants is not sufficient to ensure satisfactory development of the plants and satisfactory yields. The use of combined nitrogen (present in the soil or in the form of fertilizer) is thus essential in the cultivation of legumes. In the early stages of vegetative development, mineral nitrogen ensures a good start for the crop, and at the end of the growth cycle, when nitrogen fixation declines sharply, the supply of nitrogen may serve as a relay after the symbiosis. Neverthless, the process of nitrate assimilation inhibits symbiotic activity and is a limiting factor in the optimal exploitation of the fixation of atmospheric nitrogen.

104

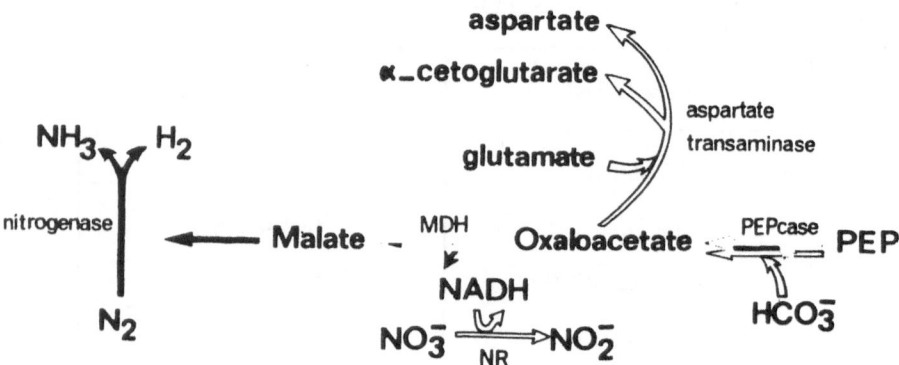

Fig. 3   Hypothetical scheme of competition between cytosolic NO_3^- reduction and bacteroids N_2 fixation. ➡ : CO_2 ⇨ : $CO_2 + NO_3^-$

Biochemical studies of this phenomenon aimed at defining the physiological criteria involved in the choice of effective symbiosis in the presence of combined nitrogen are thus indispensable. They apply to agronomic studies to determine the environmental factors that allow legumes to fix N_2 while simultaneously exploiting the combined nitrogen in the soil most efficiently and to the search for symbiotic associations capable of fixing N_2 in the presence of high concentrations of combined nitrogen. This one depends partially on variability within each species and each cultivar of legumes (Serraj, unpublished results) and also within the strains of *Rhizobium* (Mc Neil, 1982).

Thus, the exploitation of the high genetic variability in both participants in the symbiosis and the development of selection programmes that include physiological and biochemical criteria for nitrogen metabolism in legumes should result in improved nitrogen fixation by the legume-*Rhizobium* symbiosis, even in the presence of high levels of combined nitrogen.

REFERENCES

Becana, M.P., Aparicio-Tejo, P.M. and Sanchez-Diaz, M. 1985. Nitrate and nitrite reduction in the plant fraction of alfalfa root nodules. Physiol. Plantarum, 65,185-188.
Beevers, L. and Hageman, R.H. 1980. Nitrate and nitrite reduction.In " The biochemistry of plants"(Eds. P.K.Stumpf and E.E. Conn) (Academic press, New York) vol5,pp 115-168.
Bello, A.B., Ceron-Diaz, W.A., Nickell, C.D., El Sherif, E.O. and Davis, L.C. 1980. Influence of cultivar, between row spacing, and plant population of fixation of soybeans. Crop Sci., 20, 751-755.

Buttery, B.R. 1986. Effects of soil nitrate level on nitrogen distribution and remobilisation in field-grown soybeans (Glycine max (L.) Merr.).Can. J. Plant Sci., 66, 67-77.

Campbell, W.H. 1976. Separation of soybean leaf nitrate reductases by affinity chromatography. Plant Sci. Lett., 7, 239-247.

Chen, P.C. and Phillips, D.A. 1977. Induction of root nodule senescence by combined nitrogen in *Pisum sativum* L. Plant Physiol., 59, 440-442.

Drevon, J.J., Kalia, V.C., Heckmann, M.O., Tillard, P., Pedelahore, P. and Salsac, L. 1988. *In situ* open-flow assay of acetylene reduction activity by soybean root nodules : influence of acetylene and oxygen. Plant Physiol. Biochem., in press.

Duc, G. 1980. Activité fixatrice d'azote des légumineuses. Possibilités de sélection. Beaucouzé. Angers 24-26 sept .

Gibson, A.H. 1976. Recovery and compensation by nodulated legumes to environmental stress. In " Symbiotic Nitrogen Fixation" (Ed. P.S. Nutman). (Cambridge University Press, Cambridge). pp. 385-404.

Harper, J.E. 1974. Soil and symbiotic nitrogen requirements for optimum soybean production. Crop Sci., 14, 255-260.

Harper, J.E. 1975. Contribution of dinitrogen and soil fertilizer nitrogen to soybean (*Glycine max* L. merr) production. In " World Soybean Research. Proceedings of the world soybean research conference" (Ed. L.D. Hill). (Urbana) pp. 101-107.

Harper, J.E. and Cooper, R.L. 1971. Nodulation response of soybeans (*Glycine max* (L.) Merrill) to application rate and placement of combined nitrogen. Crop Sci., 11, 438-440.

Heckmann, M.O. and Drevon, J.J. 1987. Nitrate metabolism in soybean root nodules. Physiol. Plantarum, 69, 721-726.

Heichel, G.H., Barnes, D.K. and Vance, C.P. 1981. Nitrogen fixation of alfalfa in the seedling year. Crop Sci., 21, 330-335.

Houwaard, F. 1979. Effect of ammonium chloride and methionine sulfoximine on the acetylene reduction of detached root nodules of peas (*Pisum sativum*). Appl. Environ. Microbiol., 87, 73-79.

Hunter, W.J. 1983. Soybean root and nodule nitrate reductase. Physiol. Plantarum, 59, 471-475.

Keister, D.L. and Evans, H.J. 1976. Oxygen requirements for acetylen reduction by pure cultures of *Rhizobia* J. Bacteriol., 129, 149-153.

Kimou, A., Drevon, J.J. and Salsac, L. 1985. Effets de l'azote combiné sur l'efficacité relative apparente de la nitrogénase chez le soja. Physiol. Vég., 23, 249-256.

La Rue, T.A. and Patterson, T.G. 1981. How much nitrogen do legumes fix? Adv. Agron., 34, 15-39.

Lawn, R.J. and Brun, W.A. 1974. Symbiotic nitrogen fixation in soybeans : I. Effect of photosynthetic source-sink manipulations. Crop Sci., 14, 11-16.

Mague, T.H. and Burris, R.H. 1972. Reduction of acetylene and nitrogen by field-grown soybeans. New Phytol., 71, 275-286.

Minchin, F.R., Minguez, M.I., Sheehy, J.E., Witty, J.F. and Skot, L. 1986. Relationships between nitrate and oxygen supply in symbiotic nitrogen fixation by white clover. J. Exp. Bot., 37, 1103-1113.

Mc Neil, D. 1982. Variations in ability of *Rhizobium japonicum* strains to nodulate soybeans and maintain fixation in the presence of nitrate. Appl. Environ. Microbiol., 44, 647-652.

Obaton, M., Miquel, M., Robin, P., Conejero, G., Domenach, A.M. and Bardin, R. 1982. Influence du déficit hydrique sur l'activité réductrice d'acétylène et nitrogénase chez le soja (*Glycine max* L. Merr, cv Hodgson). C.R. Acad. Sci. Paris. 284, 1007-1012.

Oghoghorie, C.G.O. and Pate, J.S. 1971. The nitrate stress syndrome of the nodulated field pea (*Pisum sativum*). Plant Soil, special volume, 185-202.

Pate, J.S. 1973. Uptake, assimilation and transport of nitrogen componds by plants. Soil Biol. Chem., 5, 109-119.

Ralston, E.J. and Imsande, J. 1982. Entry of oxygen and nitrogen into intact soybean nodules. J. Exp. Bot., 33, 208-214.

Rennie, R.J. and Kemp, G.A. 1983. N_2-fixation in field beans quantified by ^{15}N isotope dilution.I. Effect of strains of *Rhizobium phaseoli*. Agron. J., 75, 640-644.

Richards, J.E. and Soper, R.J. 1978. Effect of N fertilizer on yield, protein content and symbiotic N fixation in faba beans. Agron. J., 71, 807-811.

Rigaud, J. and Puppo, A. 1977. Effect of nitrite upon leghemoglobin and interaction with nitrogen fixation. Biochim.Biophys.Acta, 497, 702-706.

Ronson, C.W. and Prinrose, S.B. 1979. Carbohydrate metabolism in *Rhizobium trifolii* : identification and symbiotic properties of mutants. J. Gen. Microbiol., 112, 77-88.

Salsac, L., Drevon, J.J., Zengbe, M., Cleyet-Marel,J.C. and Obaton, M. 1984. Energy requirement of symbiotic nitrogen fixation. Physiol. Vég., 22, 509-521.

Stephens, B.D. and Neyra, C.A. 1983. Nitrate and nitrite reduction in relation to nitrogenase activity in soybean nodules and *Rhizobium japonicum* bacteroids. Plant Physiol., 71, 731-735.

Streeter, J.G. 1982. Synthesis and accumulation of nitrite in soybean nodules supplied with nitrate. Phant Physiol., 69, 1429-1434.

Streeter, J.G. 1985 Nitrate inhibition of legume nodule growth and activity. I. Long term studies with continuous supply of nitrate. Plant Physiol., 77, 321-324.

Tjepkema, J.D. and Yocum, C.S. 1974. Measurement of oxygen partial pressure by oxygen microelectrodes. Planta, 119, 351-360.

Trinchant, J.C. and Rigaud, J. 1980. Nitrite inhibition of nitrogenase from soybean bacteroids. Arch. Microbiol., 124, 49-54.

Trinchant, J.C. and Rigaud, J. 1981. Acetylene reduction and respiration of bacteroids isolated from french-beans receiving nitrate. Physiol. Plantarum, 53, 511-517.

Trinchant, J.C., Birot, A.M. and Rigaud, J. 1981. Oxygen supply and energy yielding substrates for nitrogen fixation (acetylene reduction) by bacteroid preparations. J. Gen. Microbiol., 125, 159-165.

VARIATION IN NODULATION ABILITY BY STRAINS OF R. TRIFOLII IN THE PRESENCE OF NITRATE

P.M. Murphy

The Agricultural Institute
Johnstown Castle Research Centre
Wexford, Ireland

ABSTRACT

Strains of R. trifolii (73) were isolated from grassland soils of zero and high fertiliser N input and from a long term tillage soil. Isolates were tested for symbiotic effectivity on white clover (T. repens) growing in the absence and presence of nitrate. All soils contained R. trifolii and strains were predominantly effective on white clover with only 3 ineffective isolates detected.

Onset of nodulation in the presence of nitrate (3.5 mM initial allowed to deplete with plant growth) was studied with the above Rhizobium strains. Results indicate that tolerance to nitrate is variable and influenced both by Rhizobium strain and host plant cultivar. A number of R. trifolii isolates were identified as being particularly sensitive to nitrate. Two effective isolates, one nitrate sensitive and the other nitrate tolerant were further characterised in the absence and presence of different nitrate concentrations in an attempt to explain the contrasting nodulating abilities of these rhizobia.

INTRODUCTION

It is estimated that European white clover based pastures can fix up to 200 kg N/ha annually. Diverse factors such as host plant-Rhizobium interaction, environmental conditions and farm management practice can all act to affect the amount of N fixed. In the case of grass-clover systems special problems arise as a result of the common practice of applying N fertiliser to increase output from mixed swards. The application of 50 kg N/ha for early season grass production can reduce nitrogen fixation by 25-30% (Murphy et al.

F. O'Gara et al. (eds.), Physiological Limitations and the Genetic Improvement of Symbiotic Nitrogen Fixation, 107–116.
© 1988 by Kluwer Academic Publishers.

1986). Hoglund and Brock (1982) reported wide variations in white clover nitrogen fixation measured over a three year period which they related to soil nitrate levels.

Variable sensitivity to nitrate has been reported between legume species and some results suggest that pasture legumes may be more tolerant of nitrate than other legume species (Harper and Gibson 1984). The isolation of legume mutants tolerant to nitrate has been reported (e.g. Carroll et al. 1985). The microsymbiont has also been studied in this regard and Nutman and Ross (1969) referred to the possibility of screening for rhizobia adapted to acidity and nitrogenous fertiliser, which could have agricultural value. Variable nodulation and nitrogen fixation ability in the presence of nitrate has been reported among strains of R. japonicum (McNeil 1982) and R. leguminosarum (Nelson 1983). The present work was undertaken to determine to what extent variation in nodulation ability in the presence of nitrate occurred between strains of R. trifolii isolated from selected soils.

METHODS

Isolation of strains

R. trifolii strains were obtained from permanent pasture soils receiving (a) no fertiliser - 7 soils (b) in excess of 100 kg N/ha - 35 soils and (c) from a tillage soil which was in continuous production of malting barley for 14 years and receiving from 55 to 99 kg N/ha. Rhizobium isolates were obtained from plants growing in perlite which had been repeatedly inoculated with an aqueous suspension prepared from soil cores collected at the various sites.

Rhizobium strain evaluation

The aseptic agar tube culture method was used to evaluate strain performance in the presence and absence of nitrate. Nitrate concentration in the nutrient agar was allowed to deplete with plant growth and onset of nodulation was visually assessed using 20 replicates per determination. Strain effectiveness was based on plant fresh weight at 8 weeks from sowing. Root nodulation patterns were determined on thin layer agar plates as described by Bhuvaneswari et al. (1981).

Liquid culture of rhizobia and preparation of cell-free extracts

Rhizobia were cultured in synthetic medium containing glutamate (0.11%) as N source and carbon supply was either 0.5% mannitol or myo-inositol. Cultures (30 ml) were harvested by centrifugation and the pellet taken up in 0.5 ml 25 mM Tris pH 7.0. Chilled cells were ruptured by ultrasonic disintegration.

Enzyme assays

Nitrogenase activity of nodulated plants was determined by acetylene reduction and polyol dehydrogenase in cell-free extracts of R. trifolii as described by Primrose and Ronson (1980).

RESULTS AND DISCUSSION

Effectivity of isolates

Although some ineffective strains were observed (Fig. 1) the isolates were predominantly effective on the chosen host plant cultivar (Grasslands Huia white clover). The presence of R. trifolii in the long term tillage soil was to be expected as Nutman and Ross (1969) reported the occurrence of this organism in Rothamsted soils

110

where wheat had grown continuously for more than 100 years. It is not

possible to say if the clustering of highly effective Rhizobium types

in the tillage compared with the pasture soils is of any significance

as only one tillage soil was examined and results give no indication

of rhizobial numbers as cell counts on the soils were not performed.

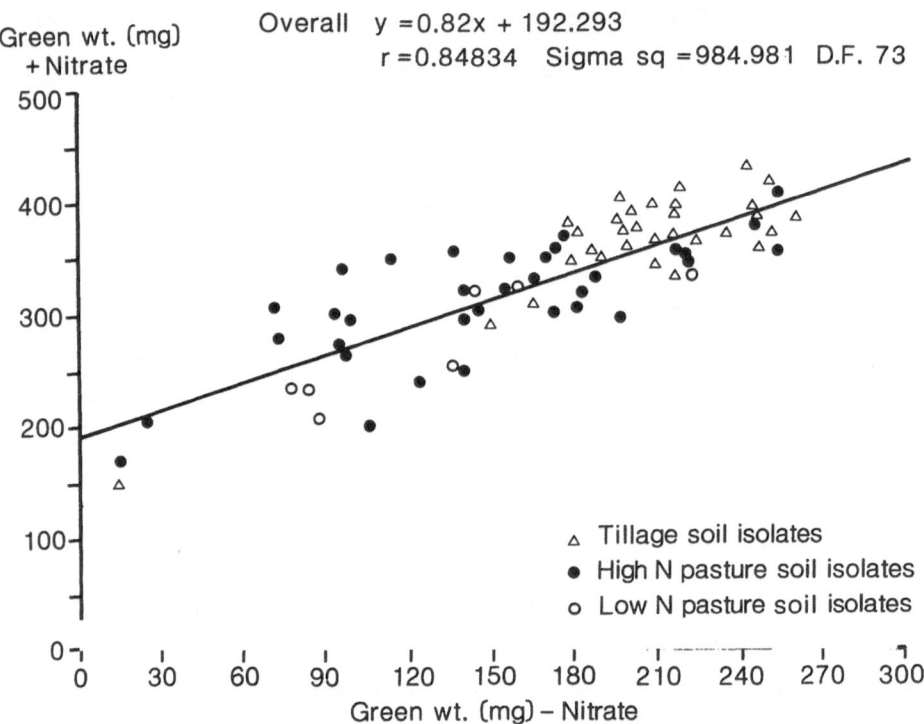

Fig. 1: Rhizobium strain effectivity in the absence and presence of
nitrate (3.5 mM initial). Values are mean of 20 replicates

Tolerance of strains to nitrate

Plant nodule emergence in the absence and presence of nitrate was

determined at time intervals following inoculation with individual

strains of R. trifolii. (Fig. 2) On day 11 following inoculation

nodulation by the strains in the absence of nitrate averaged 93%.   In contrast, percentage nodulation of the nitrate treated plants on day 11 (when nitrate levels had dropped to ca. 1.5 mM) varied from zero to 84%, indicative of a strain/nitrate interaction in the white clover symbiosis. There was no evidence of an adaptation of strains to high levels of N in the field as tolerance to nitrate was randomly distributed among isolates from high and zero N-treated soils.

This method of screening Rhizobium strains was a rapid and simple procedure although in the middle ranges differences are small and not easily reproducible. However results were consistent at the extremes of tolerance to nitrate as evidenced by the repeat analysis for strains 6 and 58. Behaviour in the presence of nitrate was a stable characteristic with these two strains although it is seen that extreme sensitivity to nitrate was an unusual phenomenon in the strains of R. trifolii examined.

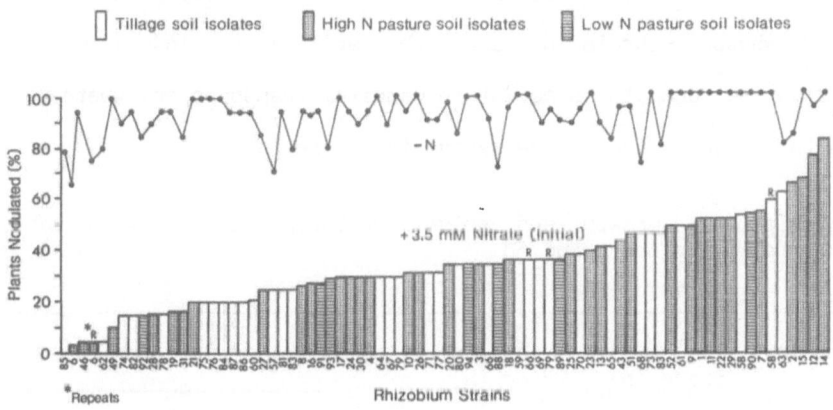

Fig. 2: Ranking of nodule emergence by Rhizobium strain on day 11 post inoculation in the absence (.____.) and presence of 3.5 mM nitrate (initial). Values are mean of 20 replicates

A variable tolerance to nitrate in strains of other Rhizobium species has also been demonstrated (Heichel and Vance, 1979; McNeil, 1982; Nelson, 1983). The mechanism whereby combined N depresses nodulation in legumes is poorly understood and likewise information is lacking on the reason for the variable susceptibility of different host plant - Rhizobium strain combinations to nitrate. With a view to obtaining further information in this area, two effective strains having contrasting nodulation patterns in the presence of nitrate were selected for further study. The nitrate sensitive strain 6 was obtained from a high N treated pasture soil, and the nitrate tolerant strain 58, came from the tillage area.

Characterisation of strains

Both strains had similar exponential growth rates though isolate 6 had an extended lag phase relative to 58 (Table 1). Strain 6 was unable to grow on mannitol. Both strains utilised myo-inositol and had similar rates of inositol dehydrogenase activity. It is not certain whether strain 6 lacks the ability to induce mannitol dehydrogenase due to a defect in mannitol transport or whether it lacks the capacity for enzyme synthesis per se.

Table 1: Growth and polyol dehydrogenase activity of R. trifolii strains 6 and 58

Strain	Lag phase (h)	Mean generation time (h)	Polyol dehydrogenase*	
			Mannitol	Myo-Inositol
6	8.0	6.0	0	218
58	5.8	5.6	109	267

*n moles NAD^+/min/mg protein

Nodulation characteristics of strains 6 and 58 are summarised in Table 2 and on the basis of plant yield in the absence of combined N, both strains had comparable effectivities. Although nitrate reduced nodulation with both strains, in the case of isolate 6 it was markedly inhibitory.

Table 2: Performance of selected R. trifolii strains 6 and 58 in the absence and presence of nitrate. Values are the mean of 30 replicates

Strain	Effectivity*	% Plants nodulated+	
	- N	- N	+ N
6	184	69	4.0
58	214	98	48

* Plant green weight (mg)8 weeks from sowing
+ 18 days post inoculation. Nitrate 8.4 mM (initial)

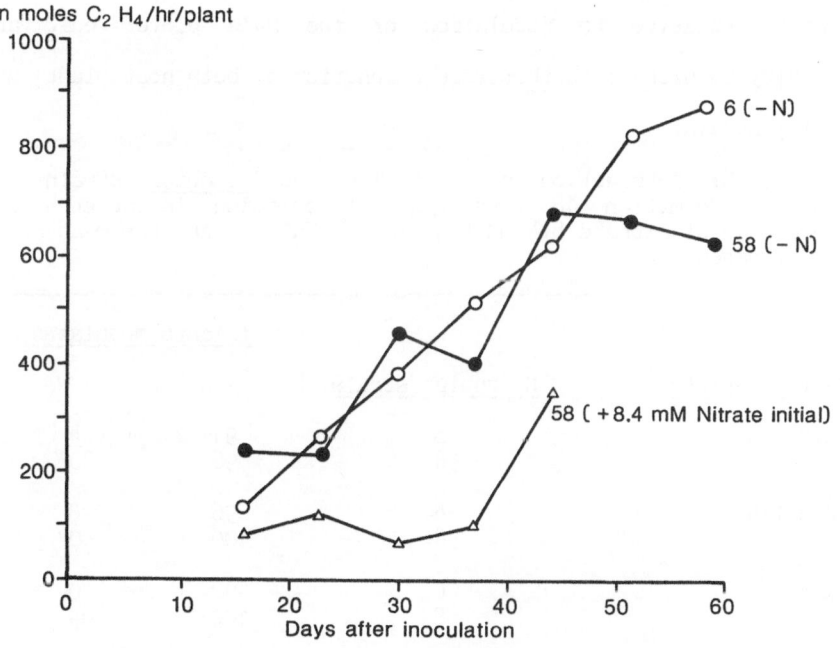

Fig. 3: Acetylene reduction by R. trifolii isolates in the absence and presence of nitrate 8.4 mM (initial)

114

In the absence of nitrate, plants inoculated with either strain had similar rates of acetylene reduction though initial activity with isolate 6 was lower due to delayed nodulation relative to strain 58 (Fig. 3). In the presence of nitrate, increasing acetylene reducing activity is evident in the case of isolate 58 (coinciding with declining nitrate concentration in the growth medium) though significant nodulation had not occurred in this case for strain 6 over the time course of the experiment.

Rhizobium strain by host plant cultivar

In the absence of N both strains showed similar nodulation on all three host plant varieties tested (Table 3). In the presence of N the sensitivity of isolate 6 to nitrate was not significantly affected by the host plant cultivar though this was not the case with strain 58 where the symbiosis in the Kentish variety showed decreased nitrate tolerance relative to nodulation of the Huia clover cultivar. Sensitivity to nitrate is therefore a function of both host plant and Rhizobium strain.

Table 3: Interaction of host plant and Rhizobium strains on nodule formation 18 days post inoculation in absence and presence of nitrate 8.4 mM (initial). Values are the mean of 20 replicates

		%Plants Nodulated	
Clover Variety	Rhizobium strain	− N	+ N
N.Z. Huia	6	91	4
	58	96	80*
Kentish	6	96	8
	58	100	50*
Aran	6	92	0
	58	100	64

*P < 0.05

Nodulation pattern

Both strains produced characteristic nodulation patterns on Huia host legume with isolate 6 typically eliciting clusters of localised large nodules and 58 forming smaller nodules scattered on the main tap and side roots. The distribution pattern has been quantified (Table 4) and it is evident that isolate 58 forms twice as many nodules as strain 6 though the latter compensates by producing larger nodules.

Table 4: Effect of Rhizobium strain on root nodule distribution in the absence and presence of nitrate, 30 days post inoculation. Nitrate 4.2 mM (initial). Values are the mean of 20 replicates

Strain	% Plants nodulated	Main root	Side root junction	Side root	Total
		Mean number nodules/plant			
6 −N	89	1.36	1.13	0.29	2.78
58 −N	97	4.33	0.45	0.48	5.26
58 +N	80	2.60	0.32	0.07	2.99
Standard Error		0.2736	0.1147	0.1088	0.2831

The host genotype (Lange and Parker, 1960) has also been shown to influence the distribution of nodules on the root system and Purchase and Nutman (1957) suggested the phenomenon of transient susceptibility of legume root hairs to infection by Rhizobium as an explanation for variable nodulation patterns. Bhuvaneswari et al. (1981) observed that nodulation in white clover occurred in two distinct phases − an initial phase confined to immature developing root hairs and a second phase involving mature root hairs. Lack of initial phase nodulation by isolate 6 may account for the characteristic nodulation pattern of this strain in the absence of N. The markedly depressive effects of N

116

on nodulation by this strain may then arise from interference by nitrate with the second phase of infections so that nodulation is further delayed.

However the mechanism whereby the different nodulation patterns arise has still to be determined. Studies presently in progress are concerned with determining binding affinities of the strains to root hairs and analysis of the Rhizobium cell surface polysaccharides.

REFERENCES

Bhuvaneswari, T.V., Bhagwat, A. and Bauer, W.D. 1981. Transient susceptibility of root cells in four common legumes to nodulation by rhizobia. Plant Physiol., 68, 1144-1149.
Carroll, B.J., McNeil, D.L., and Gresshoff, P.M. 1985. A supernodulation and nitrate-tolerant symbiotic soybean mutant. Plant Physiol. 78, 34-40.
Harper, J.E., and Gibson, A.H. 1984. Differential nodulation tolerance to nitrate among legume species. Crop. Sci., 24, 797-801.
Heichel, G.H., and Vance, C.P., 1979. Nitrate -N and Rhizobium strain roles in alfalfa seedling nodulation and growth. Crop Sci. 19, 512-518.
Hoglund, J.H., and Brock, J.L. 1982. Biological nitrogen inputs in pastures. In "Proceedings of a workshop on nitrogen balances in terrestrial ecosystems in New Zealand". Palmerston North N.Z. May 1980, 67-75.
Lange, R.T., and Parker, C.A., 1960. Nodulation patterns on legumes. Nature 186, 178-179.
McNeil, D.L. 1982. Variations in ability of R. japonicum strains to nodulate soybeans and maintain fixation in the presence of nitrate. Appl. Envir. Microbiol. 44, 647-652.
Murphy, P.M., Turner, S. and Murphy, M.E. 1986. Effect of spring applied urea and calcium ammonium nitrate on white clover (T. repens) performance in a grazed ryegrass-clover pasture. Ir. J. agric., Res., 25, 251-259.
Nelson, L.M. 1983. Variation in ability of R. leguminosarum isolates to fix dinitrogen symbiotically in the presence of ammonium nitrate. Can. J. Microbiol., 29, 1626-1633.
Nutman, P.S. and Ross, G.J.S. 1969. Rhizobium in the soils of the Rothamsted and Woburn farms. In "Rothamsted experimental station report" - part 2, 148-167.
Primrose, S.B. and Ronson, C.W. 1980. Polyol metabolism by R. trifolii. J. Bact. 141, 1109-1114.
Purchase, H.F. and Nutman, P.S. 1957. Studies on the physiology of nodule formation. V1: The influence of bacterial numbers in the rhizosphere on nodule initiation. Ann. Bot., 21, 439-454.

POSITIVE AND NETATIVE ASPECTS OF DENITRIFICATION IN

RHIZOBIUM-LEGUME ASSOCIATIONS

S. Casella

Istituto di Microbiologica Agraria,
Universita di Pisa,
Via del Borghetto, 80
56100 Pisa, Italy.

ABSTRACT

Although denitrification by Rhizobium is well documented, the implications of the porcess for both free-living cells and the Rhizobium-legume symbiosis is still unclear. Several findings suggest possible positive implications, while others highlight possible detrimental effects of this pathway on nitrogen fixation as well as on the environment. The present paper is an attempt to emphasise that in order to fully benefit from rhizobial activities in agricultural ecosystems, it is necessary to understand the denitrification process in detail.

INTRODUCTION

The dissimilatory reduction of nitrate to N_2O or N_2 by

Rhizobium spp. has recently become an important area of

research. Relatively few rhizobial strains have been tested

for denitrification activity, but many differences have

already been described as regards the energetics (Rigaud et

al., 1973; Daniel et al., 1980; Bhandari et al., 1984;

Shapleigh and Payne, 1985a), the products and intermediates

(Zablotowicz et al., 1978; Kasper and Tiedje, 1980; Daniel

et al., 1982; O'Hara et al., 1983; Shapleigh and Payne,

1985b) (see Fig. 1), the physiology (Daniel and Appleby, 1972;

O'Hara et al., 1983; Casella et al., 1986), the occurrence

(Zablotowicz et al., 1978; Daniel et al., 1982; Casella et

al., 1984a,b) and the ecological aspects of this process both

F. O'Gara et al. (eds.), Physiological Limitations and the Genetic Improvement of Symbiotic Nitrogen Fixation, 117–125.
© 1988 by Kluwer Academic Publishers.

118

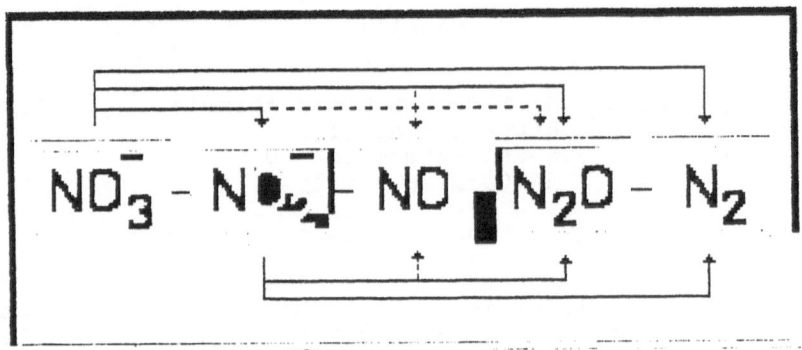

Fig. 1. End products and intermrdiates of the rhizobial denitrification pathway.

on free-living and symbiotic rhizobia (Phillips et al., 1978;
Crozat et al., 1982; O'Hara et al., 1984).

Recent findings suggest that denitrification may
represent an important advantage both for Rhizobium and
Rhizobium-legume associations because (i) it is a possible way
by which Rhizobium can survive under anaerobic conditions
using nitrate as a terminal electron acceptor (Daniel and
Appleby, 1972), (ii) energy can be obtained by proton
translocation during nitrate, nitrite, nitric oxide and
nitrous oxide respiration (Bhandari et al., 1984; Shapleigh
and Payne, 1985a), (iii) nitrogen fixation in bacteroids can
be supported by denitrification (Rigaud et al., 1973) and (iv)
given the extreme sensitivity of nitrogenase to nitrite and
nitric oxide, the ability to remove nitrite and other N-oxides
via denitrification can be considered to be a detoxification
mechanism (Trinchant and Rigaud, 1980; Castillo and Cardenas,
1982; Casella et al., 1986).

Since denitrification may occur in the field at rates comprable to those at which rhizobia fix nitrogen (O'Hara et al., 1984) the potential threat of the process is its ability to remove fixed nitrogen from the soil. The inhibitory effects that the intermediates of the process (especially nitrite and nitric oxide) have on nitrogenase and leghemoglobin (Rigaud and Puppo, 1977; Trinchant and Rigaud, 1982) and the inhibition they produce on the oxygen-uptake of free-living rhizobia are negative aspects of the denitrification process.

POSITIVE EFFECTS OF DENITRIFICATION

Under anaerobic soil conditions denitrifying capability enhances survival of rhizobia which may utilise nitrate or other N-oxides as a terminal electron acceptor. Since most slow-growing rhizobia denitrify, Crozat et al. (1982) suggested that this capabiltiy may play an important role in the observed persistence of these rhizobia in soils.

Oxygen seems to be the main factor controlling the synthesis of denitrifying enzymes in several rhizobial strains tested. This appears to be the situation irrespective of the presence or absence of nitrate (O'Hara et al., 1983; Casella et al., 1986). However, nitrate reductase activity has been detected in Rhizobium cells grown aerobically (Daniel et al., 1980).

Both oxygen and nitrate may be used as an electron acceptor for proton translocation in some strains of B. japonicum with H^+/O_2 ratio being twice that of H^+/NO_3^- (Daniels et al., 1980). Bacteroids formed by such B. japonicum strains are able to fix nitrogen under anaerobic conditions using nitrate as an electron acceptor for energy generation (Rigaud et al., 1973; Bhandari et al., 1984).

Although energy may be produced during the reduction of nitrate to nitrite, some doubts exist about the amount of energy that can be generated during the reduction of nitrite or other N-oxides. Shapleigh et al., (1985a) have reported detectable values of H^+/NO_2^- and H^+/NO ratios for B. japonicum strain USDA 6. These authors found H^+/oxidant ratios of about 1.0 on supplying the strain with NO_2^-, NO or N_2O which would be consistent with the typically low yields (about 10-20% of aerobically grown cultures) obtained with denitrifiers. As denitrification is not an efficient process in terms of energy generation, it is possible that the process serves other purposes in the cell. Nitrate reduction results in the accumulation of nitrite which is a toxic compound. In Rhizobium "hedysari" strain HCNT1 oxygen-uptake rates are inhibited by concentrations of nitrite as low as 3 μM (Casella et al., 1986). The rapid reduction of nitrite to nitrous oxide or dinitrogen results in the conversion of a toxic compound to a non-toxic gaseous product (unlikely to be reconverted to nitrite as may happen if the reduction product is ammonia). It is unlikely that the latter reaction would occur as the reduction of nitrite to N_2O requires only $2e^-$ compared to the $6e^-$ necessary for the conversion to ammonia. Therefore, a detoxification mechanism may be the real function of N-oxide reduction.

Such a mechanism may offer additional advantages since in the nodules nitrogenase may be exposed to high levels of nitrite (Streeter, 1982). Trinchant and Rigaud (1982) have reported that nitrite strongly inhibits acetylene reduction activity in bacteroids isolated from soybean nodules. Furthermore nitric oxide is also known to inhibit nitrogenase activity as well as hydrogen evolution. The presence of a complete denitrifying system in bacteroids may considerably reduce such inhibitory effects (see Fig. 2) with a consequential benefit to the nitrogen fixation process.

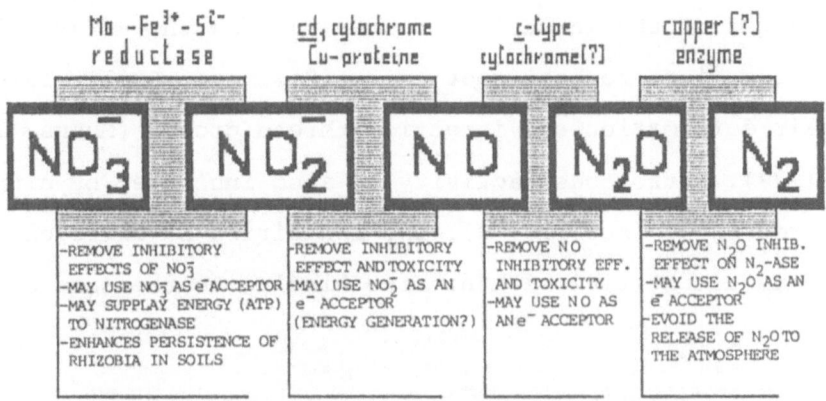

Fig. 2. Proposed benefits produced by the rhizobial denitrification system.

DETRIMENTAL EFFECTS OF DENITRIFICATION

Rhizobium lupini inoculated into the soil at a level of 10^4 cells g^{-1} soil can deplete nitrogen from the soil at a rate of 20 Kg N ha^{-1} yr^{-1} (O'Hara et al., 1984). Considering that many soils that have been examined contain more than 10^5 Rhizobium g^{-1} soil (Parker et al., 1977; O'Hara et al., 1984) rhizobial denitrification may be as important as nitrogen fixation in terms of the availability of fixed nitrogen. Betlach and Tiedje (1981) reported k values for nitrate reduction in the range 5-15 μM, and these concentrations are lower than the concentration of nitrate found in the soil solutions of even unfertilized soils. However, Phillips et al. (1978) have suggested that the kinetics of denitrification under most field conditions may be influenced more by the rate of nitrate diffusion through the soil than by its concentration in the soil solution.

The intermediates of nitrate reduction are known to affect symbiosis more intensely than nitrate itself by inhibiting the early stages of the infection process such as the attachment of the microsymbiont to host root hairs (Dazzo and Brill, 1978), root hair deformation and infection thread growth (Munns, 1977; Dart, 1977). Nitrogenase activity is also inhibited by nitrite and nitric oxide (see Fig. 3). The inhibition of nitrogenase by nitrite is competitive with nitrite binding to the same site on

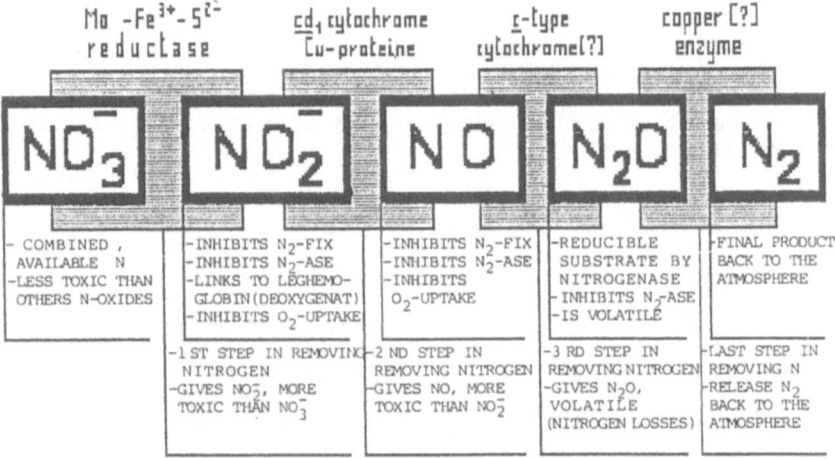

Fig. 3. Proposed detrimental effects of the rhizobial denitri-fication system.

the MoFe protein as nitrogen (Trinchant and Rigaud, 1980). The Fe protein is not affected when it is incubated alone with nitrite. Nitric oxide and nitrous oxide can also inhibit nitrogenase enzyme activity (Trinchant and Rigaud, 1982).

Changes observed in the oxyleghemoglobin spectrum after

treatment with nitrite clearly show that nitrite can react with heme resulting in a rapid deoxygenation of leghemoglobin (Rigaud and Puppo, 1977). The appearance of an absorption peak at 627 nm is consistent with the conversion of ferro-leghemoglobin to the ferric form.

CONCLUDING REMARKS

The lack of sufficient information on the role of rhizobia in the soil nitrogen budget and especially on denitrification can result in some aspects of the same phenomena being inter-preted sometimes as positive and sometimes as negative. Such ambiguous interpretations also arise as a result of the extreme versatility of rhizobia making it impossible to extrapolate denitrification results to all rhizobial strains. It is not clear for instance, why some rhizobia reduce nitrate to dinit-rogen gas and others only to nitrite, some strains are nitrate respirers only reducing nitrate to nitrite and others denitrify apparently only from nitrite, being unable to reduce nitrate at substantial rates. Moreover, some strains possess a copper containing nitrite reductase while some have a cytochrome cd reductase.

Although the capability of some bacteroids, which release nitrogen and fix it simultaneously under certain conditions appears to be anomalous, the possession of both these pathways "must" be important to rhizobia and rhizobia-legume symbiosis. What is needed is the clarification of the eventual benefit or at least the real effects of rhizobial denitrification in agri-cultural ecosystems.

124

REFERENCES

Betlach, M.R. and Tiedje, J.M. 1981. Kinetic explanation for accumulation of nitrate, nitric-oxide and nitrous oxide during bacteria denitrification. Appl. Environ. Microbiol. 42, 1074-1084.

Bhandari, B., Naïk, M.S. and Nicholas, D.J.D. 1984. ATP production coupled to the denitrification of nitrate in Rhizobium japonicum, grown in cultures and in bacteroids from Glycine max. FEMS Lett. 168, 321-326.

Burris, R.J. 1979. Inhibition. In "A Treatise on Dinitrogen Fixation", Sect. II, Biochemistry, (Eds. R.W.F. Hardy and R.G. Burns), (Wiley, New York), pp. 569-604.

Casella S., Leporini, C. and Nuti, M.P. 1984a. Nitrous oxide production by nitrogen-fixing fast-growing rhizobia. Microb. Ecol. 20, 107-114.

Casella, S., Leporini, C. and Nuti, M.P. 1984b. Denitrification by fast-growing rhizobia. In "Advances in Nitrogen Fixation Research" (Eds. C. Veeger and W.E. Newton) (Nijhoff/Junk/Pudoc, The Hague) p. 250.

Casella, S., Shapleigh, J.P. and Payne, W.J. 1986. Nitrite reduction in Rhizobium "hedysari" strain HCNT 1. Arch. Microbiol. 146, 233-238.

Castillo, F., Cardenas, J. 1982. Nitrite inhibition of bacterial dinitrogen fixation. Naturforsch 37c, 748-786.

Chen, P.C. and Phillips, D.A. 1977. Induction of root nodule senescence by combined nitrogen in Pisum sativum. Plant Physiol. 59, 440-442.

Crozat, Y., Cleyet-Marel, J.C., Giraud, J.J. and Obaton, M. 1982. Survival rates of Rhizobium japonicum populations introduced into different soils. Soil Biol. Biochem. 14, 401-405.

Crutzen, P.J. and Enhalt, D.H. 1977. Effects of nitrogen fertilizers and combustion on the stratospheric ozone layer. Ambio. 6, 112-117.

Daniel, R.M. and Appleby, C.A. 1972. Anaerobic-nitrate, symbiotic and aerobic growth of Rhizobium japonicum : effects of cytochrome P-450, other haemoproteins, nitrate and nitrite reductases. Biochim. Biophys. Acta. 275, 347-354.

Daniel, R.M., Smith, I.M., Phillips, J.A.D., Ratcliffe, H.D., Drozd, J.W. and Bull, A.T. 1980. Anaerobic growth and denitirfication by Rhizobium japonicum and other rhizobia. J. Gen. Microbiol. 120, 517-521.

Daniel, R.M., Limmer, A.W., Steele, K.W. and Smith, I.M. 1982. Anaerobic growth, nitrate reduction and denitrification in 46 Rhizobium strains. J. Gen. Microbiol. 128, 1811-1815.

Dart, P. 1977. Infection and development of leguminous nodules. In "A Treatise on Dinitrogen Fixation", Sect. II, Biology, (Eds., R.W.F. Hardy and W.S. Silver), Wiley, New York, pp367-472.

Dazzo, F.B. and Brill, W.J. 1978. Regulation by fixed nitrogen of host-symbiont recognition in the Rhizobium-clover symbiosis. Plant Physiol. 62, 18-21.

Finke, R.L., Harper, J.E. and Hageman, R.H. 1982. Efficiency of nitrogen assimilation by N2-fixing and nitrate-grown soybean plants (Glycine max L. Merr.), Plant Physiol. 70, 1178-1184.

Kasper, H.F. and Tiedje, J.M. 1980. Response of electron capture detector to hydrogen, oxygen, nitrogen, carbon dioxide, nitric oxide and nitrous oxide. J. Chromatogr. 193, 142-147.

Munns, D.N. 1977. Mineral nutrition and the legume symbiosis. In " A Treatise on Dinitrogen Fixation", Sect. IV, Agronomy and Ecology, (Eds., R.W.F. Hardy and W.S. Silver) Wiley, New York, pp. 353-392.

Noel, K.D., Carneol, M. and Brill, W.J. 1982. Nodule protein synthesis and nitrogenase activity of soybeans exposed to fixed nitrogen. Plant Physiol. 70, 1236-1241.

O'Hara, G.W., Daniel, R.M. and Steele, K.W. 1983. Effect of oxygen on the synthesis, activity and breakdown of the Rhizobium denitrification system. J. Gen. Microbiol. 129, 2405-2412.

O'Hara, G.W., Daniel, R.M., Steele, K.W. and Bonish, P.M. 1984. Nitrogen losses from soils caused by Rhizobium-dependent denitrification. Soil Biol. Biochem. 16, 429-431.

Parker, C.A., Trinick, M.J. and Chatel, D.L. 1977. Rhizobia as soil and rhizosphere inhabitants. In "A Treatise on Dinitrogen Fixation", Sect. IV, Agronomy and Ecology, (Eds., R.W.F. Hardy and A.H. Gibson), Wiley, New York, pp. 311-352.

Phillips, R.E., Reddy, K.R. and Patrick, W.H. Jr. 1978. The role of nitrate diffusion in determining of the order and rate of denitrification in flooded soil. II. Theoretical analysis and interpretation. Proc. Amer. Soil Sci. Soc. 42, 272-278.

Rigaud, J. and Puppo, A. 1977. Effect of nitrite upon leghemoglobin and interaction with nitrogen fixation. Biochim. Biophys. Acta 497, 702-706.

Rigaud, J., Bergersen, F.J., Turner, G.L. and Daniel, R.M. 1973 Nitrate dependent anaerobic acetylene-reduction and nitrogen-fixation by soybean bacteroids. J. Gen. Microbiol 77, 137-144.

Shapleigh, J.P. and Payne, W.J. 1985a. Differentiation of c, d1 cytochrome and copper nitrite reductase production in denitrifiers. FEMS Microbiol. Letts. 26, 275-279.

Shapleigh, J.P. and Payne, W.J. 1985b. Nitric oxide-dependent proton translocation in various denitrifiers. J. Bacteriol 163, 837-840.

Streeter, J.G. 1982. Synthesis and accumulation of nitrite in soybean nodules supplied with nitrite. Plant Physiol. 69, 1429-1434.

Trinchant, J.C. and Rigaud, J. 1980. Nitrite inhibition of nitrogenase from soybean bacteroids. Arch Microbiol. 124, 49-54.

Trinchant, J.C. and Rigaud, J. 1982. Nitrite and nitric oxide as inhibitors of nitrogenase from soybean bacteroids. Appl. Environ. Microbiol. 44, 1385-1388.

Zablotowicz, R.M., Eskew, D.L. and Focht, D.D. 1978. Denitrification in Rhizobium. Can. J. Microbiol. 24, 757-760.

COMPETITION BETWEEN NODULATING AND NON-NODULATING RHIZOBIUM STRAINS: DELAY OF NODULATION

T.A. Lie, G.J. Nijland[*] and S.H. Waluyo

Department of Microbiology,
Agricultural University,
Wageningen, The Netherlands.

ABSTRACT

Pea cv. Afghanistan requires highly specific Rhizobium strains which are present in soils of the Middle East (M.E.). Most European R. leguminosarum strains are unable to nodulate the Afghan pea but they can interfere with nodulation of the M.E. Rhizobium strains. A strongly competitive European strain can prevent nodulation almost completely, when applied together with the M.E. strain Tom. Other European strains are less competitive but they can interfere with nodulation of strain Tom, when applied at very high numbers. Besides reducing the number of root nodules formed, the European strains also delay the formation and functioning of the nodules induced by strain Tom. This effect can also be observed with Rhizobium strains from other cross-innoculation groups, particularly with those from the clover group.

INTRODUCTION

We reported previously that European R. leguminosarum

strains, although unable to nodulate the primitive pea cv.

Afghanistan, can compete effectively with Rhizobium strains

from the Middle East (M.E.) which are capable of nodulating

this pea (Lie et al., 1978; Winarno and Lie, 1979). In particu-

lar we studied the interaction of the Dutch Rhizobium strain PF_2,

which was selected as one of the best Rhizobium strains for

peas in The Netherlands (Wieringa and Bakhuis, 1957) and known

for its highly competitive ability (Lie et al., 1979). When

strain PF_2 was applied on the same day as strain Tom (from

---

[*]Dedicated to Jan Nijland, who died shortly after finishing his research

F. O'Gara et al. (eds.), Physiological Limitations and the Genetic Improvement of Symbiotic Nitrogen Fixation, 127–136.
© 1988 by Kluwer Academic Publishers.

Turkey) to the Afghan pea, nodulation of the latter strain was completely supressed. However, when strain PF_2 was applied 24-48 hours later than strain Tom, nodulation took place, showing that there is a critical period shortly after inoculation (Winarno and Lie, 1979). Recent studies demonstrate that the ability of strain PF_2 to block nodulation is due to genes located on a plasmid (pSym PF_2), which can be transferred to a Rhizobium strain from a completely different cross-inoculation group (Dowling et al., 1987). In this paper we will show that, besides supressing nodule number, non-nodulating Rhizobium strains can also delay the appearance of nodules.

COMPETITIVE ABILITY OF R. LEGUMINOSARUM STRAINS PRE AND PF_2.

There is a large variation in the competitive ability among Rhizobium strains, which is expressed in the number of nodules occupied by a particular strain, when applied together with other strains to the same host plant. We used two Dutch R. leguminosarum strains PRE and PF_2 , which are serologically related (unpublished results) but differ in their competitive ability. For the identification of the bacteria in the nodules, we prepared mutants resistant to 250 µg streptomycin and 50 µg acriflavin per ml. When these drug resistant mutants were applied in equal numbers with the parental strains to pea plants, about 50% of the nodules were found to be derived from the mutant strains (Table 1). These results showed that the mutants are not significantly changed from the parental strains with regard to their competitive ability.This experiment also showed that strain PF_2 is a stronger competitor than strain PRE.

TABLE 1   Distribution of root nodules, containing drug-resistant
Rhiobium strains, on roots of pea cv. Rondo, growing in a N-free
nutrient solution under aseptic conditions. The plants were either
inoculated with a single strain or with a mixture of strains PRE
and PF_2 and their drug-resistant derivatives PRE** and PF_2**

Rhizobium strain(s)	Nodules (%) from drug-resistant mutants	
	at day 16	at day 23
PRE	0	0
PRE**	100	100
PF_2	0	0
PF_2**	100	100
PRE + PRE**	40 $\pm$ 12	50 $\pm$ 14
PRE + PF_2**	98 $\pm$ 2	97 $\pm$ 4
PF_2 + PF_2**	52 $\pm$ 7	58 $\pm$ 5
PF_2 + PRE**	13 $\pm$ 9	30 $\pm$ 3

SUPPRESSION OF NODULATION IN PEA cv. AFGHANISTAN

As reported earlier (Lie, 1978) several Afghanistan peas
are unable to form root nodules with European R. leguminosarum
strains including the above mentioned Dutch strains PRE and PF_2
(Winarno and Lie, 1979). To nodulate these peas, specific
Rhizobium strains are required which are present in soils of
the Middle East and Central Asia. In our experiments a
standard strain Tom, isolated from a soil sample near Tomek,
Turkey, was used routinely. To our suprise we observed that
several European R. leguminosarum strains, although unable to
nodulate the Afghan pea, can prevent nodulation by the Turkish
strain Tom (Winarno and Lie, 1979; Table 2). In accordance
with the results in Table 1 these results show that strain PF_2
is a stronger competitor than strain PRE. Whereas no effect
was observed in the presence of strain PRE, complete suppression of nodulation was found with strain PF_2.

TABLE 2   Interaction between the nodulating Rhizobium strain Tom
and the non-nodulating strains PRE and PF_2, respectively, on nodulation of pea cv. Afghanistan, growing in a nutrient solution.

Rhizobium strain(s)	Number of Nodules per plant	Inhibition (%)
Tom	90^a	–
PRE	0^b	–
PF_2	0^b	–
Tom + PRE	101^a	0
Tom + PF_2	0^b	100

Numbers followed by the same letter are not significantly different
at the 5% level.

THE EFFECT OF BACTERIAL NUMBERS

Lie and Lommen (1979) reported that the competition
between strains Tom and PF_2 can be influenced by the numbers
of bacteria at inoculation. In our previous experiments we
routinely used equal numbers of bacteria at a standard rate of
about 10^5 _Rhizobium_ cells/ml plant nutrient solution. In the
following experiments we investigated the possibility of improving the competitive ability of a weak strain by increasing the bacterial numbers. The results in Table 3 confirmed
our earlier results that complete suppression of nodulation
can be obtained in the case of the strongly competitive strain
PF_2 with equal numbers of the two strains. With the weakly
competitive strain PRE, a 50% suppression of nodulation was
obtained with a tenfold excess of PRE and complete suppression
was achieved at a thousand-fold excess of PRE cells.

We observed frequently that the nodules formed by strain
Tom, in the presence of non-nodulating _Rhizobium_ strains, are

TABLE 3   The effect of bacterial numbers on competition between the nodulating Rhizobium strain Tom and the non-nodulating strains PRE and PF_2, respectively, on nodulation of pea cv. Afghanistan growing in a nutrient solution.

Rhizobium strain(s)	Number of Nodules per plant	Nitrogenase (nmoles C_2H_4/plant/ hour)
Tom $(0.4.10^5$/ml)	131[a]	1850
Tom $(0.4.10^5$/ml) + PRE $(10^6$/ml)	68[b]	359
Tom $(0.4.10^5$/ml) + PRE $(10^8$/ml)	6[c]	5
Tom $(0.4.10^5$/ml) + PF_2 $(2.10^5$/ml)	4[c]	8
Tom $(0.4.10^5$/ml) + PF_2 $(2.10^7$/ml)	1[c]	5

Numbers followed by the same letter are not significantly different at the 5% level

often very small and much less coloured in comparison with the nodules of the plants inoculated with strain Tom only. As shown earlier (Winarno and Lie, 1979) and in Table 3, the nitrogenase activity is also much lower. By observing the appearance of nodules daily (Fig. 1) it is evident that nodule formation is strongly delayed. The length of the delay is dependent on both the competitive ability of the strain and the number of cells used for inoculation.

THE EFFECT OF R. TRIFOLII STRAINS

So far only R. leguminosarum strains were tested on pea cv. Afghanistan. In the following section Rhizobium strains from other cross-inoculation groups and a strain of Agrobacterium tumefaciens were studied. Special attention was given to R. trifolii strains, which are closely related to R. leguminosarum strains. In order to observe any effect, these strains had to be applied at high cell numbers (approximately 10^6-10^7). In the first experiment (Table 4) live cells and killed cells

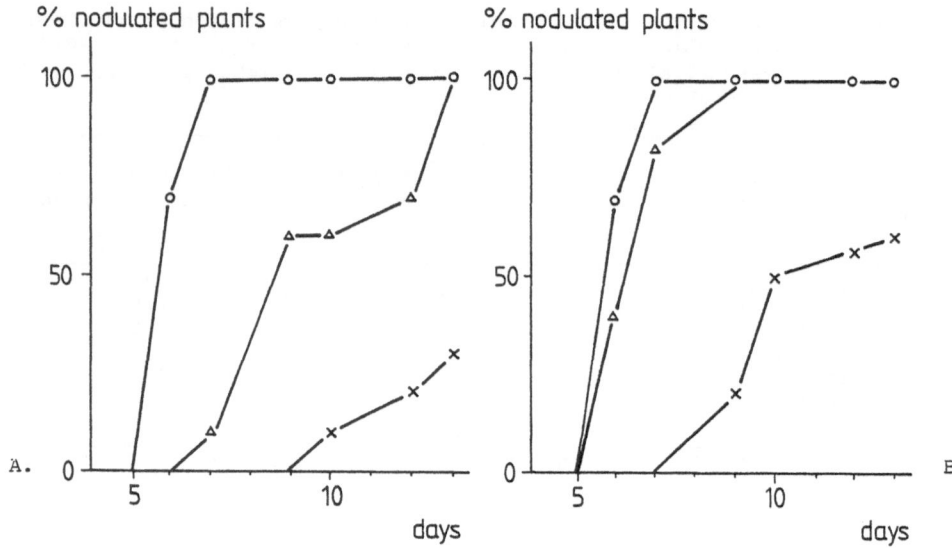

Fig. 1   Delay of nodulation in pea cv. Afghanistan in symbiosis with
the nodulating Rhizobium strain Tom only, or in the presence of the
non-nodulating strains PF$_2$(Fig.1A) and strain PRE (Fig. 1B), respectively. Only Rhizobium strain Tom (circles, 0.4.10^5/ml); with strain
PF$_2$ at a concentration of 2.10^5/ml (triangles) or 2.10^7/ml (squares);
with strain PRE at a concentration of 10^6/ml (triangles) or 10^8/ml
(squares), respectively.

TABLE 4   The effect of different Rhizobium strains and Agrobacterium
on nodulation and nitrogen fixation of pea cv. Afghanistan, inoculated
with R. leguminosarum strain Tom.

Rhizobium strain(s)	Nodule Number per plant	Nitrogenase nmoles C_2H_4/ plant/hour
R. leguminosarum Tom	54[a]	1308
Tom + R. leguminosarum PF$_2$	0.2[b]	6
Tom + R. leguminosarum PF$_2$ (killed)*	61[a]	1372
Tom + R. trifolii ClF	48[a]	981
Tom + R. phaseoli KA 47	68[a]	1133
Tom + Bradyrhizobium CB 756	50[a]	1266
Tom + Agrobacterium A8	74[a]	1461

* the cells were either killed by heat (80°C), UV light or chloroform
Numbers followed by the same letter are not significantly different
at the 5% level

of the R. leguminosarum strain PF$_2$ were used as controls.
These results show that, in comparison to R. leguminosarum
strain PF$_2$, Agrobacterium, Bradyrhizobium and Rhizobium
strains from other cross-inoculation groups had little or no
effect on nodulation by strain Tom. R. trifolii strains, in
general, do not reduce nodulation strongly, but some strains
are capable of reducing the nitrogenase activity of the
nodules (Table 5). As shown in Fig. 2 this is again due to a

TABLE 5   The effect of different Rhizobium trifolii strains on nodulation and nitrogen fixation of pea cv. Afghanistan, inoculated with
R. leguminosarum strain Tom.

Rhizobium strain(s)	Nodule Number per plant	Nitrogenase (nmoles C_2H_4/plant/hour)
R. leguminosarum Tom	89[a]	1216
Tom + R. trifolii HKC	89[a]	1168
Tom + R. trifolii ClF	68[ab]	337
Tom + R. trifolii S 459	68[ab]	419
Tom + R. trifolii S 460	59[b]	795
Tom + R. trifolii S 229	49[b]	476
Tom + R. trifolii A 121111	55[b]	397

Numbers followed by the same letter are not significantly different
at the 5% level.

delay of nodulation by the clover strain, albeit less severely
than in the presence of strain PF$_2$.

These experiments also show that live cells of strain PF$_2$
are required to suppress nodulation since cells killed by
heat, UV light or chloroform had no significant effect. This
is in agreement with the results of Broughton et al. (1982),
who used gamma irradiated cells and showed them to be normally
attached to root hairs of pea cv. Afghanistan.

134

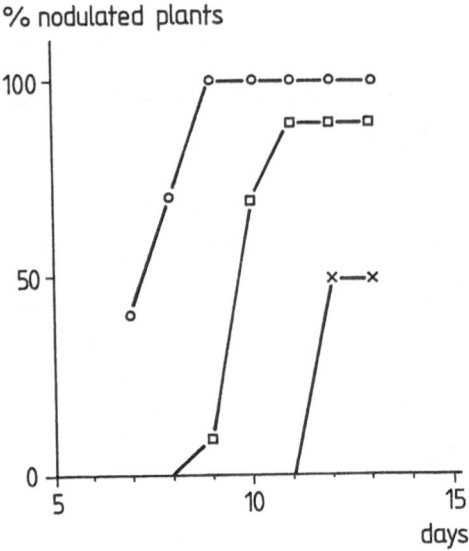

% nodulated plants

Fig 2   Delay of nodulation in pea cv. Afghanistan in symbiosis with
the nodulating R. leguminosarum strain Tom only (circles), and in the
presence of the non-nodulating R. leguminosarum strain PF_2 (crosses)
and R. trifolii strain ClF (squares).

DISCUSSION

     In our early papers (Lie et al., 1978; Winarno and Lie,
1979) we drew attention to the remarkable fact that some
Rhizobium strains unable to form root nodules on pea cv.
Afghanistan can nevertheless prevent the nodulation of another
strain. The effect is very dramatic and the suppression can be
almost 100%. Here we call attention to the effect of weakly
competitive strains, which exert only a slight effect on
nodule number. However, nitrogen fixation can be reduced
significantly and our studies show that this is mainly due to
a delay in the appearance of the root nodules.

     As far as we are aware, this delay of nodulation as a
result of competition has not been reported before. It should

be mentioned that the non-nodulating <u>Rhizobium</u> strains PRE and

PF_2 can infect the root hairs of pea cv. Afghanistan. We also

observed that several <u>R</u>. <u>trifolii</u> strains, including strain

CIF, can form infection threads in pea cv. Rondo (unpublished

results). However, we could not detect any antaganistic

effects between <u>Rhizobium</u> strain Tom and several European

<u>Rhizobium</u> strains <u>in</u> <u>vitro</u> or in the rhizosphere of pea cv.

Afghanistan.

So far we have no evidence that delay of nodulation , due

to the presence of non-nodulating <u>Rhizobium</u> strains, will also

take place in soil under natural conditions. Such a situation

may be encountered when a leguminous crop is cultivated

following another legume. The soil population of the already

present <u>Rhizobium</u> strain can be at a high level sufficient to

influence nodulation of the following plant.

REFERENCES

Broughton, W.J., Samrey, U. and Bohlool, B.B. 1982. Competi-
    tion of <u>Pisum</u> <u>sativum</u> cv. Afghanistan requires live
    rhizobia and a plant component. Can. J. Microbiol. <u>28</u>,
    162-168.
Dowling, D.N., Samrey, U., Stanley, J. and Broughton, W.J.
    1987. Cloning of <u>Rhizobium</u> <u>leguminosarum</u> genes for compe-
    titive nodulation blocking of peas. J. Bacteriol. <u>169</u>,
    1345-1348.
Lie, T.A. 1987. Symbiotic specialisation in pea plants; the
    requirement of specific <u>Rhizobium</u> strains for peas from
    Afghanistan. Ann. Appl. Biol. <u>88</u>, 462-465.
Lie, T.A., Winarno, R. and Timmermans, P.C.J.M. 1978.
    <u>Rhizobium</u> strains from wild and cultivated legumes:
    suppression of nodulation by a non-nodulating <u>Rhizobium</u>
    strain. In "Microbial Ecology" (Eds. M.W. Loutit and
    J.A.R. Miles). (Springer Verlag, Berlin, Heidelberg). pp.
    398-402.
Lie, T.A. and Lommen, W.J.M. 1979. Suppression of nodulation
    by non-nodulating <u>Rhizobium</u> strains. In "Proceedings of
    the Sixth Australian Legume Nodulation Conference, Perth"
    pp. 24-25.
Lie, T.A., Soe-Agnie, I.E., Muller, G.J.L. and Goktan, D.

1979. Environmental control of symbiotic nitrogen
fixation: Limitation to and flexibility of the legume-
Rhizobium system. In "Soil Microbiology and Plant
Nutrition" (Eds. W.J Broughton, C.K. John, J.C. Rajarao
and Beda Lim) (Malayan University Press, Kuala Lumpur).
pp. 194-212.

Wieringa, K.T. and Bakhuis, J.A. 1957. Chromatography as a
means of selecting effective strains of rhizobia. Plant
and Soil, 7, 254-262.

Winarno, R. and Lie, T.A. 1979. Competition between Rhizobium
strains in nodule formation: interaction between
nodulating and non-nodulating strains. Plant and Soil,
51, 135-142.

EFFECT OF SOIL PH ON NITROGENASE ACTIVITY, CROP YIELD AND RHIZOBIUM MULTIPLICATION.

K. Mengel and E. Schubert

Institute of Plant Nutrition of the Justus Liebig University,
6300 Giessen, Federal Republic of Germany.

ABSTRACT

In pot experiments with Vicia faba significant effects of soil pH on the nitrogenase activity and crop yield were found. Nitrogenase activity measured with a non-destructive acetylene test was closely correlated with the nitrogen fixed by the Vicia/Rhizobium symbiosis (r = 0.993). Low soil pH delayed the multiplication of Rhizobium which probably affected the infection rate of roots by the bacterium.

INTRODUCTION

Most leguminous plants require a neutral or only a slightly acid soil pH. Since leguminous plants decrease the soil pH during the growth period in soils with a poor buffer potential (Mengel and Steffens, 1982; Hauter and Steffens, 1985), it is assumed that this fall in soil pH could affect nitrogenase activity and growth of the host plant. This question was investigated in pot experiments with broad beans (Vicia faba). Besides the pH buffer potential of the soils, the effects of soil pH and exchangeable Ca^{2+} on the nitrogenase activity and on the multiplication of Rhizobium were also studied.

pH BUFFER POWER OF SOILS

The eight soils used differed in type, texture and pH buffer power (Table 1). They were limed with CaO to a neutral pH level (pH 7.00 to 7.60 measured in 0.01 M $CaCl_2$). On these

137

F. O'Gara et al. (eds.), Physiological Limitations and the Genetic Improvement of Symbiotic Nitrogen Fixation, 137–146.
© 1988 by Kluwer Academic Publishers.

soils <u>Vicia</u> <u>faba</u> (cv Kristall) was grown in pots to full
maturity. Seeds were inoculated with <u>Rhizobium</u> <u>leguminosarum</u>
before sowing. From Table 1 it is clear that the soil pH
decreased during the growth period only in soils with a lower
buffer potential. Soils with the higher buffer power (first
3 soils in Table 2) showed higher grain and N yields. In
these soils the integrated nitrogenase activity measured with
a destructive acetylene test was also higher with one
exception, the alluvium soil. The latter was rich in organic
matter (33 g kg^{-1}) and probably mineralized soil N affected
the N_2 fixation. The same could be true for the plastosol
which showed the lowest nitrogenase activity and of which the
organic matter content amounted to 26 g kg^{-1}. Omitting these
two soils the average N yield of the crop and the nitrogenase
activity of the well buffered soils amounted to 303 mmol N/pot
and 162 mmol ethylene/pot respectively, while the average N
yield and the nitrogenase activity of the poorly buffered
soils were 243 mmol N/pot and 128 mmol ethylene/pot respectively. These differences in N yield and nitrogenase activity
are significant at the 0.1% level. However, the N concentration in the upper plant parts, at the beginning of the
grain filling period, did not differ between both groups and
amounted to 2.4% and 2.5% in the dry matter for the well
buffered soils and for the poorly buffered soils respectively.
According to this finding, it is unlikely that the buffer
power of the soils directly affected nitrogenase activity.
Probably another factor, e.g. water supply, had an impact on

TABLE 1   Soil characteristics in relation to the grain yield, the yield of total N and the nitrogenase activity of Vicia faba.

Soil	Clay g kg⁻¹	Buffer pow.*	pH Initial	pH End	N yield mmol N/ pot	Grain yield g/pot	Nitro- genase**
Pararendzina	183	15430	7.60	7.56	302	74	156
Alluvium	494	7644	7.50	7.40	312	77	110
Colluvisol	173	5477	7.35	7.35	310	72	166
Brown earth	53	37	7.25	6.52	245	66	132
Plastosol	452	28	7.20	6.35	276	67	114
Grey brown pod.	187	23	7.00	6.25	247	62	144
Grey brown pod.	190	12	7.35	6.25	243	64	128
Brown earth	71	6	7.30	5.81	238	63	133

* Buffer power = $\dfrac{\Delta \text{ mol H}^+}{\Delta \text{ pH}}$

** Nitrogenase = Integrated C_2H_4 quantity in mmol/pot, calculated from the destructive acetylene tests at flowering and beginning and end of grain filling.

140

growth and CO_2 assimilation with consequences on yield production and nitrogenase activity.

SOIL pH AND EXCHANGEABLE Ca^{++}

In a further experiment, an acidic brown earth was limed with $Ca(OH)_2$ or treated with Ca salts ($CaSO_4$, $CaCl_2$) in order to establish different levels of pH and exchangeable Ca^{2+}. Vicia faba grown in pots with these treatments was tested for its nitrogenase activity with a non-destructive acetylene test at 12 dates throughout the growth period. The technique of this non-destructive acetylene method has been described else- where (Schubert, 1987). Fig. 1 shows the C_2H_4 production rates at the various dates for the pH 5.4 treatments having different levels of exchangeable Ca^{2+}. There were no major differences between the treatments. The pH 6.2 treatments with different Ca^{2+} levels had higher rates of C_2H_4

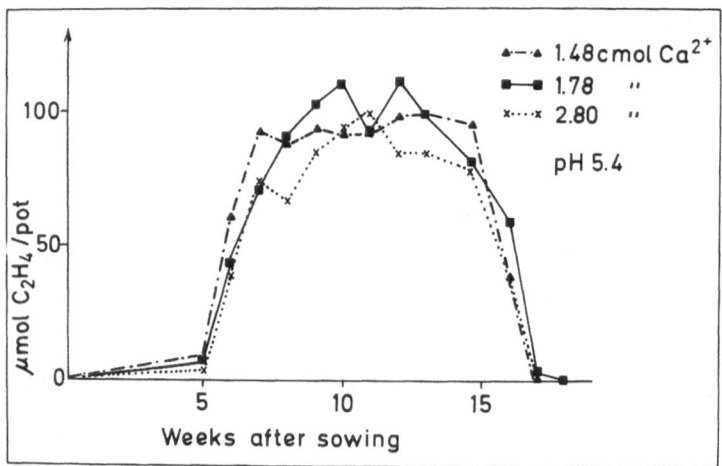

Fig. 1. Effect of exchangeable Ca^{2+} on the nitrogenase activity measured throughout the growth period with a nondestructive acetylene test.

production, but here also the level of Ca^{2+} had no significant impact on the nitrogenase activity (unpublished
results). The lowest level of exchangeable Ca^{2+} employed in
these experiments was 0.95 cmol kg^{-1} soil, which obviously was
sufficient for maximum nitrogenase activity.

In Fig. 2 the effect of soil pH on the nitrogenase
activity measured with a non-destructive acetylene test is
shown. At a pH level of 4.7 the nitrogenase activity remained
low throughout the growth period. Between the treatments pH
6.2 and pH 7.0 no major differences in nitrogenase activity
occurred. Obviously this pH range was optimum for N_2
fixation. It is evident that at the low pH levels (pH 4.7 and
5.4) the nitrogenase activity was particularly delayed at the

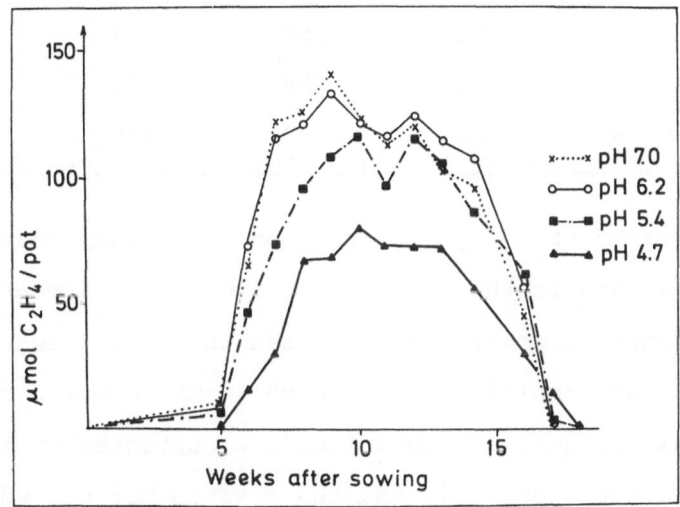

Fig. 2. Effect of soil pH on the nitrogenase activity measured
throughout the growth period with a nondestructive acetylene
test.

beginning of the growth period.

As can be seen from Table 2, the soil pH also had a clear
effect on the growth and yield of <u>Vicia</u> <u>faba</u>.

TABLE 2   Nitrogen yield in relation to the C_2H_4 production
measured by the acetylene test (E. Schubert, 1987).

Soil pH	Ca cmol kg^{-1}	Grain yield g/pot	C_2H_4 mmol/ pot	N yield mmol/ pot	C_2H_4/N
4.7	13.7	34	104	124	0.84
5.4	17.6	51	169	211	0.80
6.2	23.4	58	197	244	0.81
7.0	27.5	59	194	251	0.78
5.4	28.7	45	141	178	0.79
5.4	14.7	51	168	212	0.79
6.2	13.4	58	196	237	0.83
6.2	9.4	60	193	246	0.78

The grain yield in the pH 4.7 teatment was significantly lower
($p < 0.001$) than in the pH 5.4, 6.2 and 7.0 treatments. Also
the difference in grain yield between the pH 5.4 and 6.2
treatments was significant ($p < 0.05$). There was a clear
relationship between the amount of C_2H_4 detected by the non-
destructive acetylene test and the N yield per pot which is
more or less the N which was actually fixed by the <u>Rhizobium</u>
during the growth period. The correlation of the plot
"dectected ethylene vs N_2 fixation" was highly significant
($r = 0.993$). This finding demonstrates that the non-

destructive acetylene test used in this study is a reliable
means for assessing the N_2 fixation potential of a leguminous
plant. From Table 2 it is evident that the molar ratio
between the ethylene recovered and the N_2 fixed was almost
constant and amounted to 0.8

RHIZOBIUM MULTIPLICATION

As pointed out above in the treatments with the lower pH
values (pH 4.7 and 5.4) the nitrogenase activity was particularly affected at the beginning of the growth period. This
raised the question whether the pH had an impact on Rhizobium
multiplication. The problem was investigated with an acidic
brown earth (same soil as used for the pH and Ca experiments).
pH values were established at 4.7, 5.4 and 7.0 by additions of
$Ca(OH)_2$. The sterilized soil was inoculated with Rhizobium
leguminosarum and incubated. Soil samples were taken after
7, 13, 20 and 26 days of incubation and the number of the
Rhizobium cells was counted. Table 3 shows that multiplication of the Rhizobium was considerably influenced by soil pH.

TABLE 3. Effect of soil pH on the multiplication of Rhizobium
leguminosarum during an incubation period of 4 weeks (Counts
are per 20g soil). Data presented are the exponents to the base
10 (decadic logarithm).

Day of incub.	0	7	13	21	26
pH 4.7	4.1	4.6	5.3	6.0	5.8
pH 5.4	4.1	5.2	6.1	6.5	6.3
pH 7.0	4.1	6.2	6.5	7.2	7.1

Highest counts were obtained after 21 days of incubation and were about 15 times higher in the pH 7.0 treatment and about 3 times higher in the pH 5.4 treatment compared to the treatment with the lowest pH.

DISCUSSION

The experiments presented here have shown that the soil pH and the pH buffer power of soils have an influence on the nitrogenase activity of the Vicia faba/R. leguminosarum system. The nitrogenase activity was affected at pH levels < 6.4 with consequences on the amount of nitrogen fixed, plant growth and grain yield. Since the pH effect was particularly noticeable at the beginning of the growth period, it is suggested that low soil pH levels affect the infection of leguminous roots by bacteria. This assumption is supported by the experimental results of Oberholzer (1984) and Hohenberg and Munns (1984). Our finding that the multiplication of Rhizobium is hampered at low soil pH levels can have a direct bearing on the infection rate as low Rhizobium numbers in the soil also lead to poor infection rates. Such a relationship was found by Bauer et al. (1985) in the soybean - Bradyrhizobium japonicum system. The reduction of N to NH_3 requires 3 H electrons which is equivalent to the reduction of 1.5 C_2H_2. The actual figure of C_2H_4 detected per one N fixed, however, was only 0.8 (Table 2) which is only about half of the theoretical value. This discrepancy could be related to the fact that a longer application of acetylene hampers the nitrogenase as was found by Minchin et al. (1983). According to Witty et al. (1985)

this decline in nitrogenase activity is brought about by a
reduced O_2 supply to the bacteroid. Minchin et al., (1983)
found that the nitrogenase activity affected by acetylene was
0.4 fold of the undisturbed nitrogenase, which means that the
latter was 2.5 times higher than the acetylene affected one.
If this factor of 2.5 is applied to the C_2H_2/N ratio that we
found, namely 0.8 (Fig. 2), then a value of two is obtained. This
value is somewhat higher than the theoretical one of 1.5. However, it should be taken into consideration that a substantial
proportion of the electrons provided to the nitrogenase are consumed for the reduction of protons which leads to the production
of H_2. It, thus, appears that a value of two is a realistic
one.

In the calculation above the assumption was made that
practically no N was provided by the soil. In a parallel pot
experiment with Lolium perenne, it was shown that this particular soil (acidic brown earth) provided only a minimum quantity of N to the plant in the pH range 4.7 to 7.0.

REFERENCES

Bauer, W.D., Bhuvaneswari, T.V., Calvert, H.E., Law, I.J.,
    Malik, N.S.A. and Vesper, S.J. 1985. Recognition and
    infection by slow-growing rhizobia. In "Nitrogen
    Fixation Research Progress" (Eds. H.J. Evans, P.J.
    Bottomley and W.E. Newton) (Martinus Nijhoff, Dordrecht)
    pp. 247-253.
Hauter, R. and Steffens, D. 1985. Einfluß einer mineralischen
    und symbiontischen Stickstoffernährung auf Protonenabgabe
    der Wurzeln, Phosphataufnahme und Wurzelentwicklung von
    Rotklee. Z. Pflanzenernähr. Bodenk., 148, 633-646.
Hohenberg, J.S. and Munns, D.N. 1984. Effect of soil acidity
    factors on nodulation and growth of Vigna unguiculata in
    solution culture. Agron. J., 76, 477-481.
Mengel, K. and Steffens, D. 1982. Beziehung zwischen
    Kationen/Anionen-Aufnahme von Rotklee und Protonen-
    abscheidung der Wurzeln. Z. Pflanzenernähr. Bodenk.,

146

    <u>145</u>, 229-236.
Minchin, F.R., Witty, J.P., Sheehy, J.E. and Muller, M.A.
    1983. A major error in the acetylene reduction assay:
    Decreases in nodular nitrogenase activity under assay
    conditions. J. Exp. Bot., <u>34</u>, 631-649.
Oberholzer, H.R. 1984. Untersuchungen uber die biologische
    Stickstoffbindung bei Ackerbohnen. Mitt. Schweiz.
    Landw., <u>32</u>, 181-196.
Schubert, E. 1987. Der Einfluß des pH-Wertes und der H -
    Pufferung des Bodens auf das Wachstum und die N -
    Fixierung der Ackerbohne (Vicia faba L.) Ph.D. Thesis,
    Fachbereich 19, Justus Liebig-University, Giessen.
Witty, J.F., Skot, K. and Revsbech, W.P. 1985. Direct
    evidence for a variable barrier to diffusion into legume
    nodules. In "Nitrogen Fixation Research Progress"
    (Eds. H.J. Evans, P.J. Bottomley and W.E. Newton).
    Martinus Nijhoff, Dordrecht, p.355.

SECTION III : GENETICS OF NITROGEN FIXATION

DICARBOXYLIC ACID UTILIZATION AND NITROGEN FIXATION EFFICIENCY IN RHIZOBIUM-LEGUME SYMBIOSIS

F. O'Gara, K. Birkenhead, Y-P. Wang, C. Condon and S.S. Manian

Department of Microbiology,
University College,
Cork, Ireland.

ABSTRACT

Two separate regions of Rhizobium meliloti DNA which encode genes involved in the transport of dicarboxylic acids have been partially characterised. One region (pMC5) is approximately 27 kb long and contains sequences homologous to the Rhizobium leguminosarum dctA, B and D genes. The other region (pRK290:4:46) is approximately 40 kb long and contains no sequences homologous to the known dct genes. Plasmid pMC5 complements all R. meliloti Dct⁻ strains tested while plasmid pRK290:4:46 complements only specific Dct⁻ strains. Plasmid pRK290:4:46, but not pMC5, has the ability to increase the succinate uptake activity of both Rhizobium and Bradyrhizobium strains into which it has been mobilised. Bradyrhizobium japonicum strains containing plasmid pRK290:4:46 also exhibit an increased nitrogen fixation activity. In addition to characterising the R. meliloti dct regions, the levels of expression of nif-specific genes in R. meliloti Dct⁻ strains have also been investigated using gene fusions. The levels of expression of translational fusions of the R. meliloti symbiotic promoters P1 (nif HDK), P2 (fix ABCX) and PnifA (nifA) in bacteroids formed by Dct⁻ strains are very low compared to the levels found in bacteroids formed by the corresponding wild type strains. These results suggest that a regulatory link may exist between the utilisation of dicarboxylic acids and expression of nitrogen fixation genes.

INTRODUCTION

In the Rhizobium-legume symbiosis the high energy cost of fixing nitrogen is ultimately derived from the photosynthate supplied by the plant. Sucrose is the major source of reduced carbon entering the nodule where it is metabolised to organic acids. It is now generally accepted that the energy for nitrogen fixation is derived principally from the oxidation of organic acids. The essential role played by organic acids

149

F. O'Gara et al. (eds.), Physiological Limitations and the Genetic Improvement of Symbiotic Nitrogen Fixation, 149–157.
© 1988 by Kluwer Academic Publishers.

(especially dicarboxylic acids) during symbiosis has been highlighted by both biochemical and genetic studies. Organic acids are rapidly taken up by bacteroids (Reibach and Streeter, 1984) and support high rates of respiration and nitrogenase activity (Bergersen and Turner, 1967). Mutant strains of Rhizobium trifolii (Ronson et al., 1981), Rhizobium leguminosarum (Finan et al., 1983) and Rhizobium meliloti (Bolton et al., 1986) defective in dicarboxylic acid transport (Dct⁻) form ineffective nodules. The Dct system is a common transport system for succinate, malate and fumarate. The transport of dicarboxylic acids is an active process and requires the presence of an energised membrane (Finan et al., 1981; McAllister and Lepo, 1983). Three genes specifically required for the transport of dicarboxylic acids have been identified in R. leguminosarum. Two of these genes (dctB and dctD) encode regulatory elements while the third gene (dctA) specifies a structural transport gene (Ronson et al., 1987a). The product of the general nitrogen regulatory gene (ntrA) is also necessary for the transport of dicarboxylic acids (Ronson et al., 1987b). In our laboratory, we are interested in the role of dicarboxylic acids in the Medicago sativa - Rhizobium meliloti symbiosis. Three aspects of the system are dealt with in this communication: (1) analysis of the R. meliloti dct region, (2) exploitation of R. meliloti dct gene(s) to enhance nitrogen fixation activity and (3) regulation of nif-specific genes in R. meliloti Dct⁻ strains.

ANALYSIS OF R. MELILOTI DCT REGIONS

Two regions of R. meliloti DNA encoding genes involved in the transport of dicarboxylic acids were identified by complementation of R. meliloti Dct⁻ strains. One region (pMC5), approximately 27 kb long, is capable of complementing all our Dct⁻ mutant strains tested. Hybridization studies showed that this region contains sequences homologous to the R. leguminosarum dct A, B and D genes. A preliminary map of this region has been obtained as follows: The individual EcoRI fragments of plasmid pMC5 were subcloned into the vector pSUP102 (R. Simon, unpublished) and a restriction map for each of the EcoRI fragments was generated. From comparisons of the individual restriction maps with the restriction fragment patterns obtained on restricting pMC5, a physical map of the region was constructed. Sequences homologous to the R. leguminosarum dct genes are present on two contiguous EcoRI fragments of size 3.05 kb and 2.80 kb (Fig. 1).

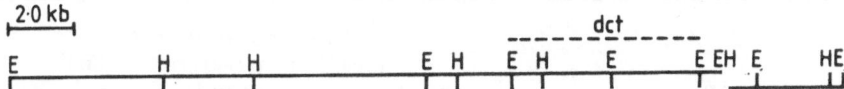

Fig. 1 : Map of the R. meliloti DNA region cloned in plasmid pMC5. The 3.05 and 2.80 kb EcoRI fragments hybridising to the R. leguminosarum dct genes are indicated. E = EcoRI, H = HindIII

The second region (pRK290:4:46), approximately 40 kb long, complements only two (CM10 and CM12) of the ten Dct⁻ strains tested (Bolton et al., 1986). Sequences homologous to the R. leguminosarum dct genes can not be detected on this region. Furthermore, hybridization studies and restriction analysis have shown that there is neither homology nor overlap between plasmids pRK290:4:46 and cosmid pMC5. This raises interesting questions as to the specific nature of the $\underline{R}$. $\underline{meliloti}$ $\underline{dct}$ gene(s) present on plasmid pRK290:4:46. We are currently attempting to characterise this $\underline{dct}$ region in order to obtain more information on the gene(s) involved.

ENHANCED SUCCINATE UPTAKE AND NITROGEN FIXATION ACTIVITIES

Initial studies carried out by Bolton $\underline{et}$ $\underline{al}$. (1986) showed that plasmid pRK290:4:46 not only complements the Dct⁻ phenotype of $\underline{R}$. $\underline{meliloti}$ CM12, but in addition increases the succinate uptake activity to levels higher than those of the parent strain CM2. Subsequent studies have demonstrated that the transfer of plasmid pRK290:4:46 into the wild type $\underline{R}$. $\underline{meliloti}$ strain as well as other $\underline{Rhizobium}$ strains such as $\underline{R}$. $\underline{leguminosarum}$ and $\underline{Bradyrhizobium}$ $\underline{japonicum}$ results in strains exhibiting increased succinate uptake activity (Table 1). The

TABLE 1

Strain	Succinate Uptake (nmol/min/mg.protein)	
	- pRK290:4:46	+ pRK290:4:46
$\underline{R}$. $\underline{meliloti}$ CM2	7.62 ± 1.60	17.19 ± 2.3
$\underline{R}$. $\underline{leguminosarum}$ CL100	26.2 ± 2.30	38.20 ± 5.0
$\underline{B}$. $\underline{japonicum}$ CJ1	1.43 ± 0.08	2.82 ± 0.15

strains containing plasmid pRK290:4:46 are capable of metabolising the extra succinate transported into the cells as evidenced by an increase in CO_2 released from these strains compared to that released from the corresponding strains lacking the plasmid. In the case of B. japonicum, the increase in the utilisation of succinate also results in an enhanced growth rate (Birkenhead et al., 1988). Plasmid pMC5 does not confer an increased succinate uptake phenotype on Rhizobium strains. While plasmid pRK290:4:46 increases the succinate uptake activity of the Rhizobium strain into which it has been mobilised, it does not appear to affect/alter the regulation of the Dct system of the host strain. The regulation of succinate uptake activity in B. japonicum CJ1 is similar to that in strain CJ1 (pRK290:4:46) with both the induced and uninduced levels being proportionately increased (Birkenhead et al., 1988). The observation that plasmid pRK290:4:46 has the capability to increase succinate uptake activity even in the absence of the normal inducer (dicarboxylic acid) is particularly interesting and focuses attention on whether the gene(s) involved are in fact conventional dct gene(s).

Several studies (Bolton et al., 1986; reviewed in Hodgeson and Stacy, 1986) have suggested that increasing the flow of dicarboxylic acids into bacteroids may result in enhanced nitrogen fixation ability. The increased succinate uptake activity of strains containing plasmid pRK290:4:46 provides an ideal system to test such a hypothesis. The maximum level of nitrogenase activity detected in strain CJ1 (pRK290:4:46) grown under microaerobic conditions is indeed substantially higher (88.6 vs 54.7 nmol C_2H_2 reduced/min/mg. protein) than that detected in strain CJ1 (Birkenhead et al., 1988). A Tn5 derivative of plasmid pRK290:4:46 unable to complement the mutant strain CM12 also lacked the ability to increase nitrogenase activity. Therefore, a single locus appears to be involved in both the increased succinate uptake and increased nitrogen fixation activities associated with plasmid pRK290:4:46. As fast-growing Rhizobium strains do not reproducibly induce nitrogenase activity under free-living conditions, the generality of the relationship between the succinate uptake and nitrogen fixation activities is currently being determined in both fast- and slow-growing rhizobial strains under symbiotic conditions.

As plasmid pRK290:4:46 affects both the efficiency of succinate uptake and the efficiency of nitrogen fixtion, it is conceivable that the plasmid could encode nifA. The gene products of dctD and nifA are very homologous and it has been suggested that the product of the nifA gene may substitute for the product of the dctD gene in activating the dicarboxylic acid transport system under certain conditions (Ronson et al., 1987a). We are currently investigating the possibility that plasmid pRK290:4:46 encodes nifA. It is unlikely that plasmid pRK290:4:46 encodes ntrA, whose gene product is now known to be required for both succinate uptake and nitrogen fixation activities, as the mutant strain CM12 complemented by plasmid pRK290:4:46 is able to grow in media containing nitrate as the sole source of nitrogen.

REGULATION OF NIF-SPECIFIC GENES

The lack of nitrogen fixation activity in Dct⁻ strains and the increased nitrogen fixation activity observed in strains exhibiting increased succinate uptake and utilisation capabilities may indicate that a co-ordinate relationship exists between carbon utilisation and nitrogen fixation. In our laboratory, we are addressing the question of how the utilisation (or lack of it) of organic acids is related to the level of nitrogen fixation observed in a particular rhizobial strain. In particular, we are interested in determining whether there is co-ordinate regulation in the transcription/ translation of nif-specific genes in response to the available energy and reductant pools? In this context, we are investigating the expression of translational fusions of the R. meliloti sym promoters P1 (nif HDK), P2 (fix ABCX) and P$_{nifA}$ (nifA) in bacteroids formed by Dct⁻ mutant strains. Plasmids pMB210 and pMB211 (Better et al., 1985) and pCHK57 (Kim et al., 1986) containing the above fusions have been mobilised into the wild type strain CM2 and the Dct⁻ strain CM12. Strain CM12 exhibits no detectable nitrogenase activity during symbiosis. β-galactosidase activity was measured in bacteroids from 3 week old nodules. The results obtained clearly showed that the levels of expression of the fusions were very low in the nodules formed by the Dct⁻ strain compared to the levels detected in nodules formed by the wild type strain (Table 2). The low levels of expression observed in the bacteroids formed by the Dct⁻ mutant strain is not as a consequence of either plasmid loss or rapid senescence of the nodule as the average percentage retention of the different plasmids in Rhizobium cells reisolated from 3 week old nodules formed by either CM2 or CM12 was 60%. Currently experiments

TABLE 2

Strain	Dct Phenotype	β-galactosidase activity (nmol/min/mg.protein)		
		pMB210 (P1)	pMB211 (P2)	pCHK57 (PnifA)
R. meliloti CM2	Dct$^+$	5065	4192	3356
R. meliloti CM12	Dct$^-$	80	106	40

are in progress to confirm the specific nature of the reduced expression of nif genes in Dct$^-$ strains.

ACKNOWLEDGEMENTS

We would like to thank Drs. Helinski and Ditta (San Diego, California), Ronson (Cambridge, Massachusetts) and Puhler, Priefer and Simon (Bielefeld, FRG) for providing various vectors, fusions and probes used in this study.

REFERENCES

Bergersen, F.J. and Turner, G.L. 1967. Nitrogen fixation by the bacteroid fraction of breis of soybean root nodules. Biochem. Biophys. Acta, 141, 507-515.
Better, M., Ditta, G. and Helinski, D.R. 1985. Deletion analysis of Rhizobium meliloti symbiotic promoters. EMBO J., 4, 2419-2424.
Birkenhead, K., Manian, S.S. and O'Gara, F. 1988. Dicarboxylic acid transport in Bradyrhizobium japonicum: Use of Rhizobium meliloti dct gene(s) to enhance nitrogen fixation. J. Bacteriol., 170 (in press).
Bolton, E., Higgission, B., Harrington, A. and O'Gara, F. 1986. Dicarboxylic acid transport in Rhizobium meliloti: Isolation of mutants and cloning of dicarboxylic acid transport genes. Arch. Microbiol., 144, 142-146.
Finan, T.M., Wood, J.M. and Jordan, D.C. 1981. Succinate transport in Rhizobium leguminosarum. J. Bacteriol., 148, 193-202.
Finan, T.M., Wood, J.M. and Jordan, D.C. 1983. Symbiotic properties of C -dicarboxylic acid transport mutants of Rhizobium leguminosarum. J. Bacteriol., 154, 1403-1413.

Hodgeson, A.L.M. and Stacey, G. 1986. Potential for Rhizobium
    improvement. CRC Crit. Rev. Biotechnol. 4, 1-73.
Kim, C.-H., Helinski, D.R. and Ditta, G. 1986. Overlapping
    transcription of the nifA regulatory gene in Rhizobium
    meliloti. Gene, 50, 141-148.
McAllister, C.F. and Lepo, J.E., 1983. Succinate transport by
    free-living forms of Rhizobium japonicum. J. Bacteriol.,
    153, 1155-1162.
Reibach, P.H. and Streeter, J.G. 1984. Evaluation of active
    versus passive uptake of metabolites by Rhizobium
    japonicum bacteriods. J. Bacteriol., 159, 47-52.
Ronson, C.W., Astwood, P.M., Nixon, B.T. and Ausubel, F.M.
    1987a. Deduced products of C -dicarboxylate transport
    regulatory genes of Rhizobium leguminosarum are homo-
    logous to nitrogen regulatory gene products. Submitted
    for publication.
Ronson, C.W., Nixon, B.T., Albright, L.M. and Ausubel, F.M.
    1987b. Rhizobium meliloti ntrA (rpoN) gene is required
    for diverse metabolic functions. J. Bacteriol., 169,
    2424-2431.
Ronson, C.W., Lyttleton, P.M. and Robertson, J.G. 1981. C -
    dicarboxylate transport mutants of Rhizobium trifolii
    form ineffective nodules on Trifolium repens. Proc.
    Natl. Acad. Sci. USA, 78, 4284-4288.

REGULATION OF NITROGEN FIXATION IN THE RHIZOBIUM MELILOTI ALFALFA SYMBIOSIS

C. W. Ronson

Biotechnica International Inc., 85 Bolton Street,
Cambridge MA 02140, USA.

ABSTRACT
    Recent results on the regulation of nitrogen fixation and C4-
dicarboxylate transport genes in Rhizobium meliloti are reviewed. nif gene
expression is activated by the nifA gene product in conjunction with the
ntrA sigma factor. Activation of the nifA gene does not require ntrC
product but occurs in response to low oxygen tension. NifA protein
activity is also sensitve to oxygen. dct regulatory gene products share
homology to nitrogen regulatory gene products and to NifA, and NtrA is
required for dct expression. The significance of the pattern of domain
conservation found between the dct, ntr, and nif regulatory proteins is
discussed.

INTRODUCTION

    The process of symbiotic nitrogen fixation takes place in root nodules
which are the result of a complex developmental process that involves the
differentiation of both the plant and bacterial partners. The mature
nodule contains plant cells filled with nitrogen-fixing rhizobia, termed
bacteroids, that are enclosed in membranes of plant origin and hence
separated from the plant cell cytoplasm. The bacteroids receive carbon
substrates, probably C4-dicarboxylates, from the plant to use as energy
substrates for nitrogen fixation, and in turn supply the plant with fixed
nitrogen in the form of ammonia. In this paper, I will review some recent
work on the genetic regulation of the nitrogen fixation (nif) genes and C4-
dicarboxylate transport (dct) genes in Rhizobium that are required for
effective nodule function. nif and dct regulation share several common
features that are also found in other regulatory systems that respond to
environmental stimuli.

REGULATION OF NIF EXPRESSION IN KLEBSIELLA PNEUMONIAE

    The structure and regulation of nif genes have been extensively studied
in Klebsiella pneumoniae, a free-living species, and these studies have
provided a model for initial studies of nif regulation in Rhizobium
meliloti, the endosymbiont of alfalfa (Medicago sativa). Many important
features of nif gene structure and regulation have been conserved in
evolution between the two species, but significant differences exist.
These differences reflect the fact that K. pneumoniae fixes nitrogen for

159

F. O'Gara et al. (eds.), Physiological Limitations and the Genetic Improvement of Symbiotic Nitrogen Fixation, 159–168.
© 1988 by Kluwer Academic Publishers.

160

its own use when sources of fixed nitrogen are exhausted, whereas
R. meliloti fixes nitrogen only in symbiosis and provides the fixed
nitrogen to the plant rather than assimilating it. Hence nitrogen fixation
by R. meliloti is attuned to the plant's needs while nitrogen fixation by
K. pneumoniae is attuned to its own needs.

In K. pneumoniae, a cluster of 17 contiguous genes, grouped into 7
operons, code for enzymes specifically involved in nitrogen fixation.
K. pneumoniae nif genes are subject to two levels of regulation in response
to ammonia and oxygen, one global and the other nif-specific. The nif-
specific level of regulation is mediated by the products of the nifLA
operon. The nifA product (NifA) is a transcriptional activator which is
required together with ntrA product (NtrA) (see below) for the expression
of all K. pneumoniae nif operons, except its own. NifL also has a
regulatory role; it appears to antagonize the action of NifA in the
presence of oxygen or intermediate levels of fixed nitrogen (reviewed by
Gussin et al., 1986).

The second level of nif regulation in K. pneumoniae is mediated by the
global system (the ntr system) that controls the expression of a variety of
nitrogen assimilatory genes in enteric bacteria. Under conditions of NH_4^+
starvation, ntrC product (NtrC) in conjunction with NtrA activates the
nifLA operon and also activates other operons involved in nitrogen
assimilation. The ntr system responds to ammonia through the action of
three additional genes, ntrB, glnD and glnB (reviewed by Kustu et al.,
1986). Under conditions of limiting ammonia, NtrB is a protein kinase that
phosphorylates NtrC to activate it (Ninfa and Magasanik. 1986). The
functional state of NtrB, in turn, is determined by the products of glnB
and glnD which respond to the intracellular 2-ketoglutarate to glutamine
ratio (Bueno et al., 1985). Thus NtrC activity is modulated by a cascade
system that is exquisitely sensitive to changes in ammonia availability.
NifA and NtrC share strong homology in their central and C-terminal domains
(Buikema et al., 1985; Drummond et al., 1986), and the promoters that they
activate do not contain canonical -35 and -10 sequences but instead have
the consensus -26 CTGGYAYR-N_4-TTGCA -10 (Gussin et al., 1986). The
requirement for NtrA for transcriptional activation by NtrC and NifA has
recently been explained by the demonstration that NtrA is a sigma factor
required to confer specificity on core RNA polymerase for the ntr promoter
sequence (Hirschmann et al., 1985; Hunt and Magasanik, 1985).

REGULATION OF NIF EXPRESSION IN RHIZOBIUM MELILOTI

Rhizobium meliloti appears to have a structural and functional
homologue of the K. pneumoniae nifA regulatory gene that is required for
nif gene expression (Szeto et al., 1984; Weber et al., 1985; Buikema et
al., 1985). Concomitant with the conservation of nifA, the structure of
nif promoters has also been conserved between K. pneumoniae and the
rhizobial species. However the mechanisms which regulate nifA
transcription or NifA activity differ between the two species.

Since Rhizobium species fix nitrogen primarily in symbiotic association
with a host legume, it seemed unlikely that the Rhizobium nif genes are
activated in response to ammonia limitation. Furthermore the R. meliloti
nifA promoter is not similar to other promoters activated by NtrC or NifA
proteins (Buikema et al., 1985), although it is strongly expressed during
symbiosis (Kim et al., 1987). On the other hand, the R. meliloti nifA gene
is directly downstream of the fixABCX operon and at least 50% of nifA mRNA
in nodules is a result of readthrough transcription from the NifA-activated
fixA promoter (Kim et al., 1987). It is therefore conceivable that the
nifH and fixA promoters are activated directly by an R. meliloti ntrC gene
in response to nitrogen limitation and that nifA transcription is
subsequently activated by its own product via the fixA promoter. To test
this idea, the R. meliloti ntrC gene was identified, on the basis of
interspecies hybridization with the E. coli ntrC gene, and the cloned gene
used to construct R. meliloti ntrC::Tn5 mutants. Unlike wild-type cells,
R. meliloti ntrC::Tn5 mutants were unable to activate the nifH or fixA
promoters in response to nitrogen limitation during free-living growth, and
were unable to use nitrate as sole nitrogen source. However, the mutants
formed effective nodules, and activated the nifH and fixA::lacZ fusions to
wild-type levels during symbiosis (Szeto et al., 1987). Szeto et al.
(1984) had previously shown that R. meliloti NifA is required for nif gene
activation during symbiotic growth but not for activation of nif promoters
ex planta. Thus, both R. meliloti NtrC and NifA appear to be able to
activate the R. meliloti nif structural gene promoters; however only NifA
is operative in the symbiotic state. The amount of NtrC activation of nifH
so far obtained during free-living growth was less than 10% of the nifA-
mediated activation obtained during symbiosis, and acetylene reduction
(nitrogen fixation) activity has not been demonstrated in free-living
R. meliloti. Hence the physiological significance of the NtrC-mediated nif
gene activation is unclear.

The observations that in alfalfa nodules NtrC is not required for
symbiotic nitrogen fixation and that the nifA promoter does not contain the
NtrA promoter consensus sequence suggest that nifA transcription initially

originates from the nifA promoter rather than the fixA promoter. This
suggestion is supported by the finding that deletions as small as 6 base
pairs at the nifA promoter cause a Fix⁻ phenotype (Ronson and Ausubel,
unpublished data). Recent experiments by Ditta et al. (1987) indicate that
the nifA promoter is activated in response to reduced oxygen tension by a
mechanism that does not require NifA or NtrC. In addition, it has recently
been found that the NifA protein from Rhizobium species is itself oxygen-
sensitive. This was first observed in experiments using NifA from
Bradyrhizobium japonicum to activate nif promoters in E. coli (Hennecke et
al., 1987). The ability of the R. meliloti NifA product to activate nif
promoters is also oxygen-sensitive, both in E. coli (J. Beynon and F.
Cannon, personal communication) and R. meliloti (Gu Qing and F. Ausubel,
personal communication). The mechanism(s) by which transcription of the
nifA gene and the activity of its product respond to reduced oxygen tension
are unknown. The response could be to molecular oxygen per se, or to some
other physiological parameter such as low redox potential. Regulation in
response to redox potential is an attractive idea, since it would provide a
mechanism to couple, through NifA synthesis and activity, the synthesis of
nitrogenase polypeptides to conditions under which they can function.
Redox control is intimately linked to carbon and oxygen supply which are
likely to be the important physiological parameters in the symbiotic
environment.

REGULATION OF DCT GENES

It is likely that the plant provides nitrogen-fixing bacteroids with
C4-dicarboxylates to use as energy substrates. The Rhizobium transport
gene for these substrates is encoded by the dctA gene, and mutants in this
gene form ineffective but structurally normal nodules. dctA is also
required by free-living rhizobia for growth on succinate and malate (Ronson
et al., 1981). The dctA promoter is strongly homologous to the consensus
promoter -26 CTGGYAYR-N4-TTGCA -10 recognized by the ntrA sigma factor
(Ronson and Astwood, 1985). Two regulatory loci dctB and dctD that are
closely linked to dctA but transcribed divergently to dctA (Ronson et al,
1987a) are required in addition to inducer for dctA activation in free-
living rhizobia. The regulatory components of the dct regulon may also be
required for symbiotic nitrogen fixation, since nitrosoguanidine-induced
dctB mutants form nodules that fail to fix nitrogen. However only one out
of 12 Tn5-induced dctB mutants formed nodules that fixed nitrogen at
significantly less than wild-type rates; the other Tn5-induced dctB mutants
and three Tn5-induced dctD mutants formed nodules that fixed nitrogen as

well as nodules formed by wild-type rhizobia (Ronson et al., manuscript in preparation). The reason for the disparity in symbiotic phenotypes between the chemically-induced and transposon-induced dctB mutants is not clear. Nevertheless, it seems likely that dctA can be activated by an alternative system, perhaps NifA, in the nodule. This suggestion is supported by the finding that DctD and NifA are structurally similar (see below).

ROLE OF NTRA IN RHIZOBIUM MELILOTI

The rhizobial NtrC and NifA proteins (Buikema et al., 1985; Szeto et al., 1987) and the promoters that they activate (Better et al., 1983) show strong homology to their enteric counterparts, suggesting that rhizobial species also contain a ntrA gene. However, such a gene has not been identified until recently, when it was found using a selection scheme based on expression of dctA. Sequence analysis of the dct regulatory genes revealed that the C-terminus of dctB gene product (DctB) is homologous to the C-terminus of NtrB and the N-terminus of dctD product (DctD) is homologous to the N-terminus of NtrC. Furthermore, DctD also shares strong homology with NtrC and NifA in a central domain of these proteins that is postulated to interact with NtrA (see below). These observations, together with the finding that dctA contains an Ntr-like promoter sequence, suggested that NtrA might be required for the expression of dctA; therefore, we formulated a selection scheme for mutations in a putative R. meliloti ntrA gene based on this supposition.

To monitor dctA expression in R. meliloti, we utilized an in-frame dctA-lacZ fusion present on a broad host range plasmid that also contained the entire dct regulon from R. leguminosarum. R. meliloti containing the dctA-lacZ fusion plasmid formed white colonies on defined medium containing 5-bromo-4-chloro-3-indolyl β-D-galactopyranoside (X-gal) with glucose as sole carbon source, and blue colonies on defined medium containing X-gal with glucose plus succinate as carbon sources. Because of this latter result and because the dctA-lacZ fusion plasmid also contained the dctB and dctD genes, we reasoned that any R. meliloti mutant that contained the dctA-lacZ fusion plasmid and formed white colonies when plated on medium containing X-gal, glucose and succinate, should be mutated in the R. meliloti ntrA gene. Three white colonies were found and all three were subsequently shown to have Tn5 inserted in ntrA as shown by DNA sequence analysis (Ronson et al., 1987b).

Consistent with the hypothesis that NtrA activates transcription in conjunction with either NtrC, DctD or NifA, R. meliloti ntrA mutants exhibited a pleiotropic phenotype: they failed to grow on nitrate as sole

164

nitrogen source and succinate as sole carbon source and they elicited Fix
nodules (Ronson et al., 1987b).   Comparisons of nucleotide sequences
surrounding the ntrA gene in R. meliloti and K. pneumoniae suggested that
genes transcribed in the same direction and possibly in the same operon as
ntrA were present upstream (ORF1) and downstream (ORF3) of the ntrA gene in
both species (Ronson et al., 1987b; unpublished data). However DNA
fragments containing intact only the R. meliloti ntrA gene were able to
complement the ntrA mutants, suggesting that the ntrA promoter lay within
125 bp of the initiation codon of the ntrA gene. Nevertheless, the strong
conservation in sequence and location of ORF1 between R. meliloti and
K. pneumoniae suggests that ORF1 may play a role in ntrA product function
(Ronson et al., 1987b). It will be interesting to determine if this is
indeed the case.

nif AND dct REGULATORY GENES ARE RELATED TO A VARIETY OF GENES THAT RESPOND
TO ENVIRONMENTAL STIMULI

As indicated above, nucleotide sequence analysis of dct, nif, and ntr
regulatory gene products revealed an unexpected degree of conservation
among these proteins. Several other bacterial proteins that are members of
two-component regulatory systems also share regions of conservation with
dctB/dctD and ntrB/ntrC (Fig. 1) (Nixon et al., 1986; Ronson et al.,
1987b,c). One component of each system is thought to act as an
environmental sensor (sensor component) that transmits a signal to the
second component (regulatory component) which then effects the response,
usually at the transcriptional level. In addition to dctB/dctD and
ntrB/ntrC, the systems include those responding to osmolarity (envZ/ompR),

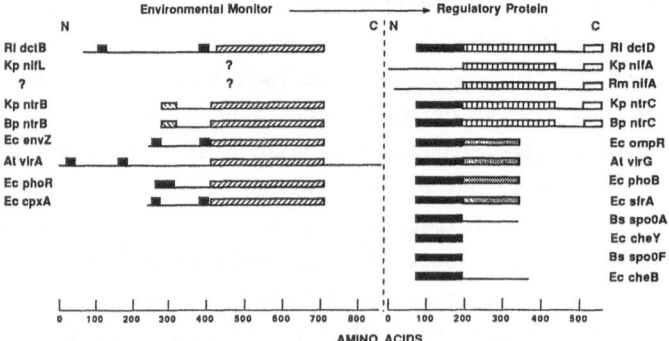

Figure 1.   Conservation of bacterial regulatory proteins that respond
to environmental stimuli. Conserved regions are shaded, and
hydrophobic regions are depicted by black boxes (see text for details).
Original references are given in Nixon et al., 1986, and Ronson et al.,
1987a,c.

phosphate limitation (phoR/phoB) and toxic compounds (changes in membrane proteins controlled by cpxA/sfrA) in E. coli, and genes controlling the virulence of Agrobacterium tumefaciens in response to plant exudate (virA/virG). About 200 amino acids of the C-terminal region is conserved among the proteins that act as sensors; about 120 amino acids of the N-terminal region is conserved among the proteins that act as regulators. The hydropathic profiles of members of the sensor class suggest that many contain an N-terminal periplasmic domain defined by two hydrophobic transmembrane regions (black boxes in Figure) and a C-terminal cytoplasmic domain, a structure similar to that of the chemosensory receptor proteins of enteric bacteria (Krikos et al., 1983). NtrB clearly does not contain a transmembrane domain; it is also the only member of the sensor class that is thought to respond to an intracellular signal (Bueno et al., 1985).

The homology among the transcriptional activators shown in the figure is not limited to the N-terminal domain; the activators also share one of two C-terminal regions. In particular, the region outside of the N terminus shared by NtrC and DctD also occurs in NifA (Buikema et al., 1985; Drummond et al., 1986).

The domain conservation in the C-terminal portions of the regulator class of proteins most likely reflects common mechanisms of transcriptional activation. A helix-turn-helix motif, characteristic of prokaryotic DNA binding proteins, is found within the C-terminal 40 amino acids of NtrC, NifA and DctD and binding sites located 100-160 base pairs upstream of the transcriptional initiation site have been identified in NtrC- and NifA-activated promoters (Hirschman et al., 1985; Reitzer and Magasanik, 1986; Buck et al., 1986). In addition, NtrC, NifA and DctD all require the alternate sigma factor NtrA (RpoN) as a co-activator, suggesting that transcriptional activation by these proteins may be mediated by a protein-protein interaction of the conserved central domain with NtrA (Ronson et al., 1987b).

The lack of conservation between the N-terminal domains of the K. pneumoniae ntrC and nifA products may account for their different sensitivities to environmental stimuli. Whereas ntrB product modulates the activity of ntrC product in response to nitrogen status, K. pneumoniae nifL product is hypothesized to inhibit nifA product activity in response to oxygen tension or low amounts of combined nitrogen (Buchanan-Wollaston and Cannon, 1984). It seems likely that in this species the N-terminal part of the nifA protein interacts with nifL protein and that the nifL protein will not contain the ntrB-set conserved sequence. Furthermore, although there is similarity between the N-terminal portions of K. pneumoniae and

R. meliloti nifA gene products, it is possible that the low degree of conservation reflects evolution of the N-terminal region of the rhizobial nifA product to respond to some aspect of the symbiotic environment (Nixon et al., 1986).

Recently, we have proposed a model for the transduction of environmental signals by the two-component regulatory systems (Nixon et al., 1986; Ronson et al., 1987c). The N-terminal domain of the environmental sensor class perceives an environmental stimulus, and transmits a signal to its conserved cytoplasmic domain through an allosteric alteration much as envisaged for the chemosensory receptors. The activated C-terminal portion of the sensor protein then interacts with and modifies the conserved N-terminal portion of the regulator protein. This modified N-terminal domain is then able to effect the response through an interaction with its C-terminal domain, the interaction causing a switch in the conformation of the C-terminus between inactive and active forms or repressor and activator forms. This model makes several predictions about the functions of the various domains of the proteins that are testable. For example, the model predicts that the N-terminal region of DctD, NtrC and NifA is regulatory. In support of this contention, we have recently found that DctD deleted of its N terminus is able to activate dctA constitutively in the presence or absence of DctB. Hence it seems that the N terminus of DctD holds the protein in inactive conformation in the absence of changes in stimuli transmitted through DctB.

ACKNOWLEDGEMENTS

I thank L. Albright, F. Ausubel, J. Beynon, F. Cannon, P. McLean, T. Nixon and W. Szeto for stimulating discussions and for sharing unpublished data.

REFERENCES

Better, M., Lewis, B., Corbin, D., Ditta, G., and Helinski, D.R., 1983, Structural relationships among Rhizobium meliloti symbiotic promoters, Cell 35:479–485.
Buchanan-Wollaston, V. and Cannon, F., 1984, Regulation of nif transcription in Klebsiella pneumoniae, in: "Advances in Nitrogen Fixation Research," C. Veeger and W.E. Newton, eds., Nijhoff/Junk, The Hague, p. 732.
Buck, M., Miller, S., Drummond, M., and Dixon, R., 1986, Upstream activator sequences are present in the promoters of nitrogen fixation genes, Nature 320:374–378.
Bueno, R., Pahel, G., and Magasanik, B., 1985, Role of glnB and glnD gene products in regulation of the glnALG operon of Escherichia coli, J. Bacteriol. 164:816–822.
Buikema, W.J., Szeto, W.W., Lemley, P.V., Orme-Johnson, W.H., and Ausubel, F.M., 1985, Nitrogen fixation specific regulatory genes of Klebsiella

pneumoniae and Rhizobium meliloti share homology with the general
nitrogen regulatory gene ntrC of K. pneumoniae, Nucleic Acids Res.
13:4539-4555.

Ditta, G., Virts, E., Palomares, A., and Kim, C., 1987, The nifA gene of
Rhizobium meliloti is oxygen regulated. J. Bacteriol. 169:3217-3223.

Drummond, M., Whitty, P., and Wootten, J., 1986, Sequence and domain
relationships of ntrC and nifA from Klebsiella pneumoniae: homologies
to other regulatory proteins, EMBO J. 5:441-447.

Gussin, G.N., Ronson, C.W., and Ausubel, F.M., 1986, Regulation of nitrogen
fixation genes, Annu. Rev. Genet 20:567-591.

Hennecke, H. et al., 1987, In "Molecular Genetics of Plant-Microbe
Interactions" (Eds. D.P.S. Verma and N. Brisson) (Martinus Nijhoff,
Dordrecht). pp 191-196.

Hirschmann, J., Wong, P.-K., Sei, K., Keener, J., and Kustu, S., 1985,
Products of nitrogen regulatory genes ntrA and ntrC of enteric bacteria
activate glnA transcription in vitro: evidence that the ntrA product is
a sigma factor, Proc. Natl. Acad. Sci. USA 82:7525-7529.

Hunt, T.P., and Magasanik, B., 1985, Transcription of glnA by purified
Escherichia coli components: core RNA polymerase and the products of
glnF, glnG, and glnL, Proc. Natl. Acad. Sci. USA 82:8453-8457.

Kim, C., Helinski, D.R., and Ditta, G., 1987, Overlapping regulatonof the
nifA regulatory gene in Rhizobium meliloti. Gene 50:141-148.

Krikos, A., Mutoh, N., Boyd, A., and Simon, M.I., 1983, Sensory transducers
of E. coli are composed of discrete structural and functional domains,
Cell 33:615-622.

Kustu, S., Sei, K.,and Keener, J., Nitrogen regulation in enteric
bacteria., In "Regulation of gene expression - 25 years on" (Eds. I.
Booth and C. Higgins). (Cambridge University Press, Cambridge) pp. 139-
154..

Ninfa, A.J. and Magasanik, B., 1986, Covalent modification of the glnG
product NR_I, by the glnL product NR_{II}, regulates the transcription of
the glnALG operon in Escherichia coli, Proc. Natl. Acad. Sci. USA
83:5909-5913.

Nixon, B.T., Ronson, C.W., and Ausubel, F.M., 1986, Two-component
regulatory systems responsive to environmental stimuli share strongly
conserved domains with the nitrogen assimilation regulatory genes ntrB
and ntrC, Proc. Natl. Acad. Sci. USA 83:7850-7854.

Reitzer, L.J. and Magasanik, B., 1986, Transcription of glnA in E. coli is
stimulated by activator bound to sites far from the promoter, Cell
45:785-792.

Ronson, C.W. and Astwood, P.M., 1985, Genes involved in the carbon
metabolism of bacteroids, in: "Nitrogen Fixation Research Progress,"
P.J. Bottomley and W.E. Newton, eds., Martinus Nijhoff, Dordrecht, pp.
201-207.

Ronson, C.W., Lyttleton, P.M., and Robertson, J.G., 1981, C4-dicarboxylate
transport mutants of Rhizobium trifolii form ineffective nodules on
Trifolium repens. Proc. Natl. Acad. Sci. USA 78:4284-4288.

Ronson, C.W., Astwood, P.M., Nixon, B.T., and Ausubel, F.M. 1987a,
submitted for publication.

Ronson, C.W., Nixon, B.T., Albright, L.M., and Ausubel, F.M., 1987b,
Rhizobium meliloti ntrA (rpoN) gene is required for diverse metabolic
functions, J. Bacteriol. 169:2424-2431.

Ronson, C.W., Nixon, B.T., and Ausubel, F.M., 1987c, Conserved domains in
bacterial regulatory proteins that respond to environmental stimuli,
Cell 49:579-581.

Szeto, W.W., Zimmerman, J.L., Sundaresan, V., and Ausubel, F.M., 1984, A
Rhizobium meliloti symbiotic regulatory gene. Cell 36:1035-1043.

Szeto, W.W., Nixon, B.T., Ronson, C.W., and Ausubel, F.M., 1987,
Identification and characterization of the Rhizobium meliloti ntrC
gene: R. meliloti has separate regulatory pathways for activation of

nitrogen fixation genes in free-living and symbiotic cells. J.
Bacteriol. 169:1423-1432.
Weber, G., Reilander, H., and Puhler, A., 1985, Mapping and expression of a
regulatory nitrogen fixation gene (fixD) of Rhizobium meliloti. EMBO J.
4:2751-2756.

STRUCTURE AND REGULATION OF FIX GENES FROM RHIZOBIUM MELILOTI

D. Kahn, J. Batut, M.L. Daveran, M. David and P. Boistard
Laboratoire de biologie moléculaire
des relations plantes-microorganismes
CNRS-INRA, BP 27, F31326 Castanet-Tolosan Cedex, France

ABSTRACT

Genetic analysis of a *fix* cluster from *Rhizobium meliloti* indicates it extends over 12.5 kb and consists of two *fix* operons on either side of a *fix* region duplicated elsewhere on the symbiotic plasmid pSym. Sequence analysis of the *fixGHIY* operon indicates it encodes a membrane-located cation-pump possibly coupled to a redox process important for symbiotic nitrogen fixation. Regulation studies demonstrate that the duplicated *fix* region is induced in symbiosis and in free-living microaerobic cultures. Induction depends on two newly described regulatory genes, *fixLJ*, but not on *nifA*, the classical *nif* activator. *fixLJ* also regulate *nif* genes in symbiosis *via* induction of *nifA* at the P*nifA* promoter. Therefore *fixLJ* regulate most presently characterized nitrogen fixation genes in *R. meliloti*: (1) *nifA*-independent *fix* genes such as *fixN* and (2) *nif* genes, by a cascade regulation involving *nifA*. The sequence of *fixLJ* shows that the FixL and FixJ proteins belong to a family of two-component regulatory systems widely spread among prokaryotes and sensitive to the external environment. This suggests that *fixLJ* mediate environmental regulation of *nif* and *fix* genes in *R. meliloti*.

INTRODUCTION

Because of their intracellular location, *Rhizobium* bacteroids may be considered as plant organelles (Verma and Long, 1983). Therefore biochemical studies aimed at the understanding of the physiology of bacteroids inside the nodule present some of the difficulties which hamper the study of organelle physiology. One of the difficulties is that the physiology of the organelle depends very tightly on the physiology of its host cell. Indeed results obtained with purified bacteroids do not necessarily reflect the actual situation inside the plant cell. Another difficulty is that it is not possible to grow isolated organelles nor differentiated bacteroids. However bacteroids differ from *stricto sensu* organelles in that they are the differentiated forms of genuine bacteria that can grow as undifferentiated cells in pure culture. Therefore *Rhizobium* bacteroids provide a unique example of an organelle amenable to genetic analysis outside the endosymbiotic context.

Mutations in genes which code for proteins known to be present in bacteroids may allow to identify whether these proteins play a role in the symbiotic nitrogen fixing association. Alternatively genetic analysis

F. O'Gara et al. (eds.), Physiological Limitations and the Genetic Improvement of Symbiotic Nitrogen Fixation, 169–174.
© 1988 by Kluwer Academic Publishers.

followed by DNA sequencing may help to identify either regulatory or structural genes. This approach will then guide further biochemical studies on the precise physiological role of the identified gene products.

We have described a cluster of *fix* genes on the pSym megaplasmid of *Rhizobium meliloti* (Batut et al., 1985ab, Renalier et al., 1987, Kahn et al., 1987) distinct from the *nif* cluster studied extensively by other laboratories (Ruvkun et al., 1982, Corbin et al., 1983, Weber et al., 1985b). Sequence analysis was used to predict possible functions for some of the genes of this cluster. Moreover we have studied the regulation of these genes and found a new regulatory circuit which also controls the expression of *nif* genes.

NITROGEN FIXATION GENES FROM RHIZOBIUM MELILOTI
The nif cluster.

Homologues of *Klebsiella pneumoniae nif* genes have been located in a 15-kb cluster on the megaplasmid pSym of *R. meliloti* (Ruvkun and Ausubel, 1980, Banfalvi et al., 1981, Rosenberg et al., 1981). The structural genes for nitrogenase form one operon *nifHDK* (Ruvkun et al., 1982). A divergent operon contiguous to *nifHDK* contains four coding sequences *fixABCX* whose protein products bear no homology with any of the *K. pneumoniae nif* gene products (Earl et al., 1987). However homologous sequences have been found in *Azotobacter vinelandii* and in *Bradyrhizobium* species and in the latter they are required for nitrogen fixation *ex planta* (Gubler and Hennecke, 1986). It has been suggested that *fixABCX* may code for electron carriers to nitrogenase.

Next to *fixABCX* has been identified a homologue of the regulatory *K. pneumoniae nifA* gene whose product is needed for the activation of the *nif* operons (Szeto et al., 1984, Weber et al., 1985a). Downstream *nifA*, *nifB* codes for a protein involved in the synthesis of the molybdenum iron cofactor of nitrogenase (Buikema et al., 1987). Another gene involved in the synthesis of this cofactor, *nifE*, has been mapped directly downstream *nifK* (Weber et al., 1985b, Hennecke et al., 1985). Another *fix* gene, *fixF*, maps close to *nodD1* (Aguilar et al., 1985) and may be functionally related to the *nif* cluster since it undergoes the *nif*-characteristic regulation by *nifA*.

The fix cluster.

To investigate the presence of additional symbiotic genes on pSym we cloned a 290 kb fragment of pSym on the broad host range vector RP4 (Julliot et al., 1984). This fragment carries the *nif* and the *nod* clusters. We have identified another symbiotic cluster on this fragment 220 kb from the *nifH* promoter (Batut et al., 1985ab). It contains two *fix* operons on either side of a 5 kb *fix* region duplicated elsewhere on pSym (Renalier et al., 1987).

In order to make predictions on possible functions of the protein products we made the complete sequence analysis of the 12.5 kb DNA which constitutes this *fix* cluster. The sequence of one operon has been analyzed in detail because it has been shown to be highly conserved among fast growing *Rhizobia* (Batut et al., 1985b, Kahn et al., 1987, Kahn D, Daveran ML, David M, Domergue O, Ghai J and Batut J, in preparation). It comprises 4 open reading frames *fixGHIY*(1). An interesting feature, the presence of an ATGA overlap of stop and start codons between adjacent genes in the operon, could be an indication of translational coupling of *fixGHIY*. All four gene products contain transmembrane sequences. FixG contains two CysxxCysxxCysxxxCys clusters typical of iron-sulfur centers and is predicted to be a redox protein. FixI is homologous with P-type ATPases (Pedersen and Carafoli, 1987), particularly with the KdpB K⁺-pump from *Escherichia coli* (Hesse et al., 1984), and is predicted to be a cation-pump involved in symbiotic nitrogen fixation. We propose that FixG, FixH, FixI and FixY may constitute a membrane-bound complex coupling the FixI cation-pump with a redox reaction catalyzed by FixG.

REGULATION OF NIF AND FIX GENES
Regulation of the nif cluster.

The *K. pneumoniae nifA* gene product is an essential activator of *nif* operons (for review see Dixon, 1984, and Gussin et al., 1986). *nifA* itself is activated under nitrogen deficiency by the *ntrC* gene product, which regulates nitrogen catabolic genes. Activity of NtrC is modulated by a second regulatory protein NtrB according to the nitrogen status of the cell (Ninfa and Magasanik, 1986). Similarly activity of NifA is modulated by a second *nif*-specific regulator NifL in the presence of oxygen or of low amounts of fixed nitrogen (Merrick et al., 1982). Thus *nif* genes are

1. *fixY* was previously named *fixX* (Kahn et al., 1987) but had to be renamed after the description of the *fixABCX* operon (Earl et al., 1987).

tightly regulated by a cascade system responding to nitrogen at a general level *via ntrBC* and to nitrogen and oxygen at a specific level *via nifLA*.

A homologue of *nifA*, but not of *nifL*, has been found in *R. meliloti* (Szeto et al., 1984, Weber et al., 1985a). As in *K. pneumoniae* it activates the other *nif* operons, namely *nifHDKE*, *fixABCX*, *nifB* and *fixF*. However symbiotic expression of *nifA* is independent of *ntrC* in *R. meliloti* since *ntrC* mutants are Fix$^+$ (Szeto et al., 1987). This is consistent with the absence of a recognizable *ntr*-dependent promoter sequence upstream *nifA* (Buikema et al., 1985). Recently Ditta et al. (1987) have shown that *nifA* can be induced in microaerobiosis *ex planta* independently of *ntrC*.

fixLJ.

Because the reiterated *fix* region is strongly expressed inside the plant (David et al., 1987) we were interested in studying its regulation. This was achieved by studying the presence of messenger RNA homologous to the DNA of the reiterated region. We also used a translational *lac* fusion with *fixN*, one of the genes of the reiterated *fix* region.

Expression of this region was not affected by a mutation in *nifA* (David et al., 1987) but by a new regulatory locus, *fixLJ* (David M, Batut J, Ghai J, Boistard P and Kahn D, in preparation). In addition we have now demonstrated that *fixLJ* are key genes of *nif* regulation in *R. meliloti* since *fixLJ* are required for *nifA* expression from its own promoter during symbiosis. Therefore *R. meliloti nif* genes are subject to a cascade regulation by which *fixLJ* activate *nifA* which in turn activates the other *nif* operons. This is very similar to *nif* regulation in *K. pneumoniae* except that *fixLJ* substitute for *ntrBC* in their *nif*-activating function. This can be related to the absence of regulation of *nif* genes by combined nitrogen in *R. meliloti* (Ditta et al., 1987).

The sequence of *fixLJ* (Daveran ML, Hertig C, Dedieu A, Domergue O and Kahn D, in preparation) shows that FixL and FixJ belong to a family of two-component regulatory systems widely spread among prokaryotes and responsive to the cell environment (Nixon et al., 1986, Ronson et al., 1987). FixL has features of a transmembrane protein and its C-terminal region shares homology with the C-terminal domains of NtrB from *Bradyrhizobium* sp. RP501, *R. meliloti* and *K. pneumoniae*, EnvZ, CpxA, PhoR and PhoM from *Escherichia coli*, VirA from *Agrobacterium tumefaciens* and DctB from *Rhizobium leguminosarum*. These proteins define the "sensor class" of Ronson et al. (1987). The N-terminal region of FixJ is homologous with the N-terminal domains of NtrC from *Bradyrhizobium*, *R. meliloti* and *K. pneumoniae*, OmpR, SfrA and PhoB from *E. coli*, VirG from *A. tumefaciens*, DctD from *R. leguminosarum*, CheB and CheY from *Salmonella typhimurium*, and SpoOA and SpoOF from *Bacillus subtilis* (Drummond et al., 1986, the "regulator class", Ronson et al., 1987). According to the current model proteins of the sensor class recognize an environmental stimulus and transmit a signal to proteins of the regulator class *via* interaction of the conserved domains. This interaction modulates the activity of the regulators perhaps by covalent modification as has been shown for NtrC (Ninfa and Magasanik, 1986). Therefore we propose that FixL senses a symbiotic signal and transduces it to FixJ, which in turn activates transcription of *nif* and *fix* genes.

Regulation by oxygen.

The expression of *fixN* could be induced in pure culture in microaerophilic conditions and this expression was found to be dependent on *fixLJ*. Similarly Ditta et al. (1987) have recently demonstrated that *nifA* can be fully induced in free-living microaerobic cultures. If microaerobic induction of *nifA* depends on *fixLJ* as does symbiotic induction, one hypothesis would be that FixL could sense the oxygen concentration and mediate oxygen regulation of *nif* and *fix* genes. Alternatively oxygen regulation could be mediated by another regulatory system that would act in conjunction with *fixLJ*. In the latter hypothesis FixL would respond to another symbiotic signal not yet identified.

REFERENCES

Aguilar OM, Kapp D and Pühler A (1985) J. Bacteriol. **164**, 245-254.
Banfalvi Z, Sakanyan V, Koncz C, Kiss A, Dusha I and Kondorosi A (1981) Mol. Gen. Genet. **184**, 318-325.
Batut J, Terzaghi B, Ghérardi M, Huguet M, Terzaghi E, Garnerone AM, Boistard P and Huguet T (1985a) Mol. Gen. Genet. **199**, 232-239.
Batut J, Boistard P, Debellé F, Dénarié J, Ghai J, Huguet T, Infante D, Martinez E, Rosenberg C, Vasse J and Truchet G (1985b) in Evans HJ, Bottomley PJ and Newton WE (eds), Nitrogen Fixation Research Progress, pp 109-115, Martinus Nijhoff, Dordrecht.
Buikema WJ, Szeto WW, Lemley PV, Orme-Johnson WH and Ausubel FM (1985) Nucleic Acids Res. **13**, 4539-4555.
Buikema WJ, Klingensmith JA, Gibbons SL and Ausubel FM (1987) J. Bacteriol. **169**, 1120-1126.
Corbin D, Barran L and Ditta G (1983) Proc. natl. Acad. Sci. USA **80**, 3005-3009.
David M, Domergue O, Pognonec P and Kahn D (1987) J. Bacteriol. **169**, 2239-2244.
Ditta G, Virts E, Palomares A and Kim CH (1987) J. Bacteriol. **169**, 3217-3223.

174

Dixon RA (1984) J. Gen. Microbiol. **130**, 2745-2755.
Drummond M, Whitty P and Wootton J (1986) EMBO J. **5**, 441-447
Earl CD, Ronson CW and Ausubel FM (1987) J. Bacteriol. **169**, 1127-1136.
Gubler M and Hennecke H (1986) FEBS Lett. **200**, 186-192
Gussin GN, Ronson CW and Ausubel FM (1986) Ann. Rev. Genet. 20, 567-591.
Hennecke H, Alvarez-Morales A, Betancourt-Alvarez M, Ebeling S, Filser M, Fischer HM, Gubler M, Hahn M, Kaluza K, Lamb JW, Meyer L, Regensburger B, Studer D and Weber J (1985) in Evans HJ, Bottomley PJ and Newton WE (eds), Nitrogen Fixation Research Progress, pp 157-163, Martinus Nijhoff, Dordrecht.
Hesse J, Wieczorek L, Altendorf K, Reicin AS, Dorus E and Epstein W (1984) Proc. Natl. Acad. Sci. USA **81**, 4746-4750.
Julliot JS, Dusha I, Renalier MH, Terzaghi B, Garnerone AM and Boistard P (1984) Mol. Gen. Genet. **193**, 17-26.
Kahn D, Batut J, Boistard P, Daveran ML, David M, Domergue O, Garnerone AM, Ghai J, Hertig C, Infante D and Renalier MH (1987) in Verma DPS and Brisson N (eds), Molecular Genetics of Plant-Microbe Interactions, pp 258-263, Martinus Nijhoff, Dordrecht.
Merrick M, Hill S, Hennecke H, Hahn M, Dixon R and Kennedy C (1982) Mol. Gen. Genet. **185**, 75-81.
Ninfa AJ and Magasanik B (1986) Proc. Natl. Acad. Sci. USA **83**, 5909-5913.
Nixon BT, Ronson CW and Ausubel FM (1986) Proc. Natl. Acad. Sci. USA **83**, 7850-7854.
Pedersen PL and Carafoli E (1987) Trends Biochem. Sci. **12**, 186-189.
Renalier MH, Batut J, Ghai J, Terzaghi B, Ghérardi M, David M, Garnerone AM, Vasse J, Truchet G, Huguet T and Boistard P (1987) J. Bacteriol. **169**, 2231-2238.
Ronson CW, Nixon BT and Ausubel FM (1987) Cell **49**, 579-581.
Rosenberg C, Boistard P, Dénarié J and Casse-Delbart F (1981) Mol. Gen. Genet. **184**, 326-333
Ruvkun GB and Ausubel FM (1980) Proc. Natl. Acad. Sci. USA **77**, 191-195.
Ruvkun GB, Sundaresan V and Ausubel FM (1982) Cell **29**, 551-559.
Szeto WW, Zimmerman JL, Sundaresan V and Ausubel FM (1984) Cell **36**, 1035-1043.
Szeto WW, Nixon BT, Ronson CW and Ausubel FM (1987) J. Bacteriol. **169**, 1423-1432.
Verma DPS and Long S (1983) Int. Rev. Cytol. **14**, 211-245.
Weber G, Reiländer H and Pühler A (1985a) EMBO J. **4**, 2751-2756.
Weber G, Aguilar OM, Gronemeier B, Reiländer H, and Pühler A (1985b) in Szalay AA, and Legocki RP (eds), Advances in the Molecular Genetics of the Bacteria-Plant Interaction, pp 13-15, Media Services, Cornell University, Ithaca, N.Y.

ORGANIZATION AND EXPRESSION OF HYDROGEN-UPTAKE (hup) GENES
OF RHIZOBIUM LEGUMINOSARUM

J. Palacios[1], A. Leyva[1], G. Ditta[2], T. Ruiz-Argüeso[1]

[1] Departamento de Microbiología
Escuela Técnica Superior de Ingenieros Agrónomos
Universidad Politécnica de Madrid
28040 Madrid, Spain.
[2] Department of Biology, B022
University of California, San Diego
La Jolla, CA 92093 USA.

ABSTRACT

Genes involved in H_2-uptake (hup genes) in Rhizobium leguminosarum strain UPM791 are clustered in a DNA region of the symbiotic plasmid. A fragment of 20.7 kilobases from this region has previously been cloned in cosmid pAL618. Analysis of 25 Tn5 insertions generated by marker exchange into the UPM791 genome indicates that at least 18 kilobases of insert DNA in pAL618 are essential for H_2-uptake in R. leguminosarum. Transposon Tn3-HoHo1, which contains lacZ gene (Stachel et al. 1985. EMBO J. 4, 891-898) was used to create lac fusions within the hup region of pAL618. High β-galactosidase activity was induced in symbiosis with peas in six Tn3-HoHo1-generated lac fusions. The corresponding Tn3-HoHo1 insertions were located in hup-specific DNA and all of them contained the lacZ gene in the left to right orientation. Two of these insertions mapped in a region showing homology to the structural genes of the B. japonicum hydrogenase, and the other four mapped in a region located outside and to the right of the putative structural region. These last four fusions also induce β-galactosidase activity in free-living cells incubated under low-oxygen conditions. These results demonstrate that we have created lac fusions within hup genes cloned in pAL618, and that these genes are expressed in a leftward to rightward orientation. It is also evident from this work that some genes located in pAL618 insert DNA are inducible in free-living cells and may be involved in regulation of hup genes expression.

INTRODUCTION

Certain strains of Bradyrhizobium japonicum, Bradyrhizobium sp. (Vigna) and Rhizobium leguminosarum have been shown to recycle the H_2 generated during nitrogen fixation in legume nodules by inducing an H_2-uptake system (Eisbrenner and Evans, 1983; Brewin, 1984). The recycling of H_2 by this system seems to have a potential for improving nitrogen fixation and productivity of legume plants (Evans et al, 1985).

Genes involved in H_2-uptake (hup genes) are located in the chromosome in B. japonicum and have been isolated and characterized (Cantrell et al, 1983; Haugland et al, 1984). In R. leguminosarum hup genes are linked to symbiotic determinants on a plasmid (Brewin et al, 1980; Nelson et al, 1985;

F. O'Gara et al. (eds.), Physiological Limitations and the Genetic Improvement of Symbiotic Nitrogen Fixation, 175–181.
© 1988 by Kluwer Academic Publishers.

Leyva et al, 1987a). By using cloned hup DNA from B. japonicum as hybridization probe, we have isolated a cosmid (pAL618) which contains genes
essential for H_2-uptake in R. leguminosarum strain UPM791 in a 20.7 kb DNA
fragment (Leyva et al, 1987b). To better understand the specific functions
required for H_2-uptake in pea nodules, we undertook a genetic analysis of
DNA cloned in pAL618. Here we present preliminary results of experiments
conducted to define transcriptional units and to study the regulation of hup
genes expression in R. leguminosarum.

MATERIALS AND METHODS

Site-directed transposon mutagenesis was carried out essentially as
described by Ditta (1986). Cosmid pAL618 (Leyva et al, 1987b) containing the
hup region of R. leguminosarum strain UPM791 was used as the target plasmid
DNA for the Tn5 insertions. Plasmid pPH1JI was used as the incoming incompa-
tible plasmid for marker exchange. Tn5 insertions into the genome were confirmed by the presence of kanamycin resistance (encoded by Tn5) and the loss
of tetracycline resistance (encoded by pAL618).

The hup-lac fusions were generated by "in vivo" inserting the Tn3-HoHo1
system, constructed by Stachel et al (1985), into DNA cloned in pAL618. Tn3-
lac insertions were selected essentially as described (Stachel et al, 1985).
β-galactosidase activity was determined in cells grown under aerobic and microaerobic conditions as described by Ditta et al (1987), and in pea bacteroids as described by Szeto et al (1984). β-galactosidase activity units
were calculated based on optical density (OD_{600}) of cultures or on bacteroid
protein content.

Pea nodule bacteroids were obtained from inoculated pea plants (Pisum
sativum L. cv. Frisson) grown either in tubes or in Leonard jar units
(Vincent, 1970). The Hup phenotype was assessed in crushed nodules measuring
H_2-dependent methylene blue reduction (Lambert et al, 1985), or in bacteroid
suspension measuring O_2-dependent H_2-uptake (Ruiz-Argüeso et al, 1978). Triparental matings were conducted as described by Ditta et al (1980) using
pRK2073 as helper plasmid. DNA manipulation techniques were standard (Maniatis et al, 1982).

RESULTS AND DISCUSSION

Organization of the hup DNA region

In order to define transcriptional units within the R. leguminosarum

hup DNA region, we used site-directed mutagenesis of DNA cloned in pAL618.
Twenty-five Tn5 transposon insertions were physically mapped, and introduced
into the UPM791 genome by homologous recombination. The resulting marker-
exchanged strains were used as inocula for pea plants and their Hup phenotype was determined in bacteroids or in crushed nodules. The location and
the effect on hydrogenase activity of Tn5 insertions are shown in Fig.1. The
Hup+ phenotype of the wild-type strain was abolished by all insertions examined except insertion 8, which mapped on the left side of pAL618 insert
DNA, and insertion 91, located at 1.5 kb from the right-hand end. These results indicate that most of the insert DNA of pAL618 is involved in H_2 oxidation. Although no positive Tn5 insertions were found in the approximately

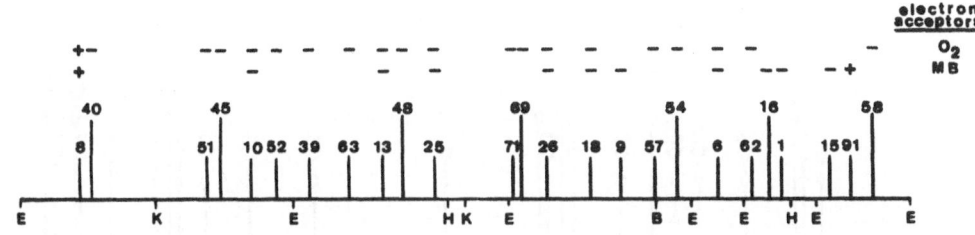

Fig. 1. Site directed Tn5 mutagenesis of pAL618 insert DNA.
Vertical lines above the restriction map indicate the position of Tn5
insertions, which are designated by numbers. The plus or minus signs
above the numbers indicate whether the insertions produced a Hup+ or
Hup⁻ phenotype when they were incorporated into the UPM791 genome by
homologous recombination. The Hup phenotype was determined by measuring
H_2-uptake with O_2 or methylene blue (MB) as electron acceptors. Restriction sites are: E (EcoRI), H (HindIII), K (KpnI) and B (BamHI).

18 kb DNA region defined by insertions 8 and 91, preliminary complementation
experiments (data not shown) indicate that at least three transcriptional
units are contained in this region. To precisely define transcriptional units,
we are currently carrying out complementation analysis of the Tn5-generated
Hup⁻ mutants with the Tn3-HoHo1 constructs described below. The Hup⁻ phenotype associated with insertion 58 suggests that a transcriptional unit starts
near the right-hand end of pAL618. This unit could extend into the adjacent
DNA, present in cosmid pAL704 (Leyva et al, 1987b), since a region with homology to B. japonicum hup DNA overlapping both cosmids has been identified
(Leyva et al, 1987b).

Direction of transcription and expression of hup genes

The determination of the direction of transcription of hup genes cloned
in pAL618 and the study of the regulation of its expression was accomplished
by using transposon Tn3-HoHo1 to construct hup-lac fusions. In this type of
constructs the expression of lacZ gene, encoded in the Tn3-HoHo1 element, is
placed under the control of the promoter of the gene into which the transposon has inserted. The expression of β-galactosidase activity only occurs
when both the reading frame of lacZ gene and the direction of transcription
of the target gene are in the same orientation.

Cosmid pAL618 was mutagenized with Tn3-HoHo1, and 37 pAL618 derived
cosmids, containing the transposon inserted into the cloned DNA, were examined. The location of the insertion and the orientation of the lacZ gene was

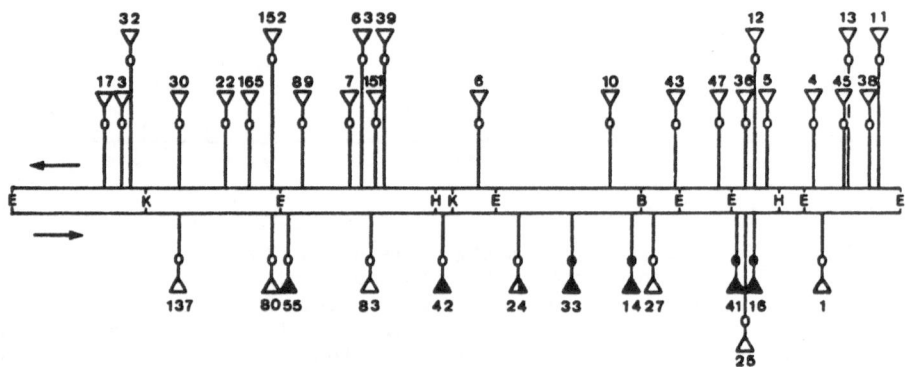

Fig.2. Expression and direction of transcription of Rhizobium leguminosarum hup genes. The expression of hup genes was monitored by measuring
β-galactosidase activity in strain UPM791 containing pAL618 derivatives
with lac fusions created by Tn3-HoHo1 insertion. Vertical lines indicate the location of Tn3-HoHo1 insertions in the insert DNA of cosmid
pAL618. Insertions above the restriction map contained the lacZ gene
in a right to left orientation, and those below the map have the lacZ
gene in the opposite direction as indicated by the horizontal arrows.
The β-galactosidase activity of lacZ gene fusions in the merodiploid
R. leguminosarum strains was determined in free-living, microaerobical-
ly grown cells (circles) and in pea bacteroids (triangles). Open, half-
solid and solid symbols indicate low, intermediate and high levels of
β-galactosidase activity (see text for details). Restriction enzyme abbreviations are as in Fig.1.

determined in each construct by restriction enzyme analysis with KpnI, BamHI
and EcoRI, enzymes for which restriction sites are asimmetrically located in
Tn3-HoHo1. Twenty-four insertions contained the lacZ gene in a right to left
orientation and the remaining 13 in the opposite direction (Fig. 2). The

pAL618::Tn3-HoHo1 constructs were individually transferred, by triparental mating, into the R. leguminosarum strain UPM791 and the resulting merodiploid strains were assayed for β-galactosidase activity in both free living and symbiotic states.

Low β-galactosidase activity (<20 U.) was observed in free-living cells of the lac fusion-containing R. leguminosarum strains grown in normal aerobic conditions, regardless of the site or orientation of the insertions. This activity was slightly higher than the endogenous β-galactosidase activity shown by UPM791(pAL618), and was considered as the basal level of activity in free-living cells. This result was expected, since hydrogenase is not constitutively expressed in cultures of Hup⁺ strains of root nodule bacteria.

When the R. leguminosarum derivative strains containing the Tn3-lac insertions were used as inocula for pea plants, nodule bacteroids from six strains showed high levels of β-galactosidase activity (>80 U.). The lacZ gene in the Tn3-HoHo1 insertions of these six strains was in a left to right orientation (Fig. 2). Low levels of activity (<35 U., basal level in symbiosis) were shown by the remaining strains, including all strains containing lac insertions in the right to left orientation. The insertions in the six strains which induced β-galactosidase activity in symbiosis are likely to be associated with hup genes, since all of them are located near Tn5 insertions leading to a Hup⁻ phenotype (Fig. 1.). Insertions 42 and 55 are located in a DNA region containing the homology to the structural genes of the hydrogenase of B. japonicum 122DES (Leyva et al, 1987b). From these results we conclude that the lacZ gene fusions created by Tn3-HoHo1 insertions 55, 42, 33, 14, 41 and 16 are under the control of specific hup gene promoters, and that these genes are transcribed in the same left to right direction. The hydrogenase structural genes of B. japonicum 122DES seems to be also transcribed in the left to right orientation in cosmid pHU52 (Zuber et al, 1986). The low level of β-galactosidase activity associated with insertions 137, 80, 83, 27, 25 and 1 (also oriented left to right) could be due to either some defect in the construct (for instance, wrong reading frame) or location of the insertions in sites outside the control of promoters expressed in symbiosis. If we assume the last possibility, then at least four transcriptional units expressed in symbiosis would be present in pAL618.

In free-living Hup⁺ strains of B. japonicum, H_2-uptake hydrogenase activity is induced when cells are incubated under an atmosphere containing H_2, CO_2 and low O_2 tensions (Maier et al, 1979). Other symbiotic genes are also induced by low oxygen concentrations in free living cells (Ditta et al,

1987). In R. leguminosarum, cultural conditions have not been found for derepression of hydrogenase activity, which seems to be expressed only in nodules. In order to test if some of the hup genes are expressed in response
to microaerobic conditions, we examined the induction of β-galactosidase activity in UPM791 derivatives containing the pAL618::Tn3-HoHo1 constructs
when cells were exposed to 1% oxygen. Four strains, containing insertions
14, 33, 41 and 16 (Fig. 2), responded to the low oxygen treatment, showing
rates of β-galactosidase activity 4 to 15 times higher than basal level.
These four insertions mapped in an approximately 4 kb DNA region located in
pAL618 to the right of the putative region of the hydrogenase structural
genes (Leyva et al, 1987b). Since all of them are also associated with expression of high β-galactosidase activity in nodules, they may define a regulatory region which responds to the low oxygen tension within the nodule,
and participate in triggering the symbiotic expression of hydrogenase structural genes. Besides microaerobic conditions, other unknown factors must be
required for the derepression of hydrogenase structural genes in free-living
R. leguminosarum. The hup-lac fusions created in this work will hopefully
allow us to study the regulation of R. leguminosarum hup genes in free-
living and symbiotic conditions.

ACKNOWLEDGMENTS

We thank S.E. Stachel for providing the Tn3-HoHo1 system. This work was
supported by grants from the CAICYT (Project nº 2462/83) and US-Spain Joint
Committee for Scientific and Technological Cooperation (Project nºCCB8509006)

REFERENCES

Brewin, N.J. 1984. Hydrogenase and energy efficiency in nitrogen fixing sym-
    bionts. In: Genes Involved in Microbe-Plant Interactions" (Ed. D.P.S.
    Verma and T.H. Hohn). (Springer-Verlag, New york) pp. 179-203.
Brewin, N.J., De Jong, T.M., Phillips, D.A. and Johnston, A.W.B. 1980. Co-
    transfer of determinants for hydrogenase activity and nodulation abili-
    ty in Rhizobium leguminosarum. Nature (London) 288, 77-79.
Cantrell, M.A., Haugland, R.A. and Evans, H.J. 1983. Construction of a Rhi-
    zobium japonicum gene bank and use in isolation of a hydrogen uptake
    gene. Proc. Natl. Acad. Sci. USA 80, 181-185.
Ditta, G., Stanfield, S., Corbin, D. and Helinski, D.R. 1980. Broad host
    range DNA cloning system for Gram-negative bacteria: construction of a
    gene bank of Rhizobium meliloti. Proc.Natl.Acad. Sci. USA 77, 7347-7351
Ditta, G. 1986. Tn5 mapping of Rhizobium nitrogen fixation genes. Methods
    Enzymol. 118, 519-528.
Ditta, G., Virts, E., Palomares, A. and Kim, C.H. 1987. The nifA gene is
    oxygen regulated. J. Bacteriol. 169, 3217-3223.
Eisbrenner, G. and Evans, H.J. 1983. Aspects of hydrogen metabolism in

nitrogen fixing legumes and other plant-microbe associations. Annu. Rev. Plant Physiol. 34, 105-136.

Evans, H.J., Hanus, F.J., Haugland, R.A., Cantrell, M.A., Xu, L.-S., Russell, S.A., Lambert, G.R and Harker, A.R. 1985. Hydrogen recycling in nodules affects nitrogen fixation and growth of soybeans. In: "World Soybean Research Conference III: Proceedings" (Ed. R. Shibles). (Westview Press, London). pp. 935-942.

Haugland, R.A., Cantrell, M.A., Beaty, J.S., Hanus, F.J., Russell, S.A. and Evans, H.J. 1984. Characterization of Rhizobium japonicum hydrogen uptake genes. J. Bacteriol. 159, 1006-1012.

Lambert, G.R., Hanus, F.J., Russell, S.A. and Evans, H.J. 1985. Determination of the hydrogenase status of individual nodules by a methylene blue reduction assay. Appl. Environ. Microbiol. 50, 537-539.

Leyva, A., Palacios, J.M. and Ruiz-Argüeso, T. 1987a. Conserved plasmid hydrogen uptake (hup)-specific sequences within Hup+ Rhizobium leguminosarum strains. Appl. Environ. Microbiol. 53, 2539-2543.

Leyva, A., Palacios, J. M., Mozo, T. and Ruiz-Argüeso,T. 1987b. Cloning and characterization of hydrogen uptake genes from Rhizobium leguminosarum. J. Bacteriol., 169, 4929-4934.

Maier, R.J., Hanus, F.J. and Evans, H.J. 1979. Regulation of hydrogenase in Rhizobium japonicum. J. Bacteriol. 137, 824-829.

Maniatis, T., Fritsch, E.F. and Sambrook, J. 1982. Molecular cloning: a laboratory manual. Cold Spring Harbor Laboratory, Cold Spring Harbor, New York.

Nelson, L.M., Grosskopf, E., Tichy, H.V. and Lotz, W. 1985. Characterization of hup-specific DNA in Rhizobium leguminosarum strains of different origin. FEMS Microbiol. Lett. 30, 53-58.

Ruiz-Argüeso, T., Hanus, F.J. and Evans, H.J. 1978. Hydrogen production and uptake by pea nodules as affected by strains of Rhizobium leguminosarum. Arch. Microbiol. 116, 113-118.

Stachel, S.E., An, G., Flores, C.and Nester, E.W. 1985. A Tn3-lacZ transposon for the random generation of β-galactosidase gene fusions: application to the analysis of gene expression in Agrobacterium. Embo J. 4, 891-898.

Szeto, W.W., Zimmerman, J.L., Sundaresan, V. and Ausubel, F.M. 1984. A Rhizobium meliloti symbiotic regulatory gene. Cell 36, 1035-1043.

Vincent, J.M., 1970. A manual for the practical study of the root nodule bacteria. International Biological Programme Handbook 15. Blackwell, Oxford.

Zuber, M., Harker, A.R., Sultana, M.A. and Evans, H.J. 1986. Cloning and expression of Bradyrhizobium japonicum uptake hydrogenase structural genes in Escherichia coli. Proc. Natl. Acad. Sci. USA 83, 7668-7672.

TWO CLASSES OF *RHIZOBIUM* GENES REQUIRED FOR THE NODULATION OF LEGUMES

A.W.B. Johnston, G.F. Hong*, D. Borthakur**, J.L. Burn and J.W. Latchford

Institute of Plant Science Research, John Innes Institute,
Colney Lane, Norwich NR4 7UH, England
*Institute of Biochemistry, Shanghai, China
**Dept. Cell Biology and Molecular Genetics, University of Chicago,
Chicago, Illinois, U.S.A.

ABSTRACT

Two different classes of genes in the bacterium *Rhizobium leguminosarum* biovar *viciae*, both of which are essential for the nodulation of its host legumes were identified. One group of genes is on a large sym plasmid and is involved in the early stages of the infection process and in the determination of host-range specificity. Several *nod* genes on the sym plasmid are not transcribed in cells growing in normal growth media but, when exposed to root exudate of legumes, are expressed at high levels. The inducer molecules in the exudate are certain flavones and flavanones and it was shown that their induction required the regulatory gene *nodD*. In *R.l.* bv. *viciae*, which nodulates peas, *nodD* is also autoregulatory, being capable of repressing its own expression. Mutant forms of *nodD* altered in their regulatory properties were isolated by *in vitro* mutagenesis of *nodD* with hydroxylamine. One class of mutation abolished both autoregulation and the induction of other *nod* genes in the presence of inducer flavonoids. Other mutations specifically inhibited the ability of *nodD* to autoregulate and a third mutant type was unaffected in autoregulation but was defective in activation of the other *nod* genes. A fourth group of *nodD* mutations, which activated transcription of other *nod* genes in the absence of inducer flavonoids was isolated; these "constitutive" forms of *nodD* caused a reduction in the number of nodules on peas and the nodules that formed failed to fix nitrogen. Using the gel retardation assay, it was shown that the *nodD* gene product bound to a DNA fragment lieing upstream of *nodD* and that this binding is probably responsible for the autoregulatory properties of *nodD*. In addition to genes on the sym plasmid, other genes, involved in the synthesis of the high molecular weight acidic exopolysaccharide [EPS], are needed for nodule formation on peas. Mutations which abolished EPS production resulted in strains which made non-mucoid colonies and which failed to nodulate. When peas were co-inoculated with two different types of Nod⁻ mutants [ie. one strain lacked its sym plasmid and the other was defective in EPS synthesis] normal numbers of nodules were formed and, in all cases, the nodules were occupied by the strain lacking its sym plasmid. Thus, a strain that does not contain a sym plasmid was "helped" into the nodule by a strain that contains such a plasmid but which is Nod⁻ because of its failure to make EPS.

183

F. O'Gara et al. (eds.), Physiological Limitations and the Genetic Improvement of Symbiotic Nitrogen Fixation, 183–192.
© *1988 by Kluwer Academic Publishers.*

INTRODUCTION

The symbiotic nitrogen-fixing interaction between legumes and the
Gram-negative bacteria that comprise the rhizobia represents one of the
most complex relationships between pro- and eukaryotes. In addition to its
major economic importance, this interaction is thus of inherent interest
since it provides a challenge in understanding how the presence of a
relatively simple bacterium in the rhizosphere can elicit a series of
defined morphological and biochemical responses in the host plant in which
genes that are normally not expressed are activated and, in parallel, the
bacteria also undergo a profound change in their behaviour, changing from
their normal rod-shaped appearance into the large, differentiated nitrogen-
fixing bacteroid forms.

For the analysis of the *Rhizobium* genes involved in the interaction
with legumes, two rather different approaches have been adopted. In one,
mutants, defective in some charecter with a readily scorable phenotype in
free-living culture are isolated and then, secondarily, the symbiotic
behaviour of such mutants is tested. The advantage of this method is that
if the mutant strain is affected in nodulation or nitrogen fixation
ability, there is a clear indication of the biochemical basis of the
defect; however, by definition, only those mutations which affect a "free-
living" phenotype will be detected. The other approach is to isolate
mutants directly on the basis of their symbiotic phenotype and then to
charecterize the function of the gene that has been so identified.
Examples of both approaches will be presented in this paper.

The species *R. leguminosarum* comprises three biovars, which are defined
and distinguished by the host legumes that they nodulate.   *R. l.* bv.
phaseoli nodulates *Phaseolus* beans, strains of bv. *trifolii* nodulate clover
and the hosts of bv. *viciae* are peas, vetches and lentils. The particular
host-range specificity of each of these biovars is determined by genes on a
large indigenous sym plasmid which, when transferred to strains of
Rhizobium with a different host-range, or indeed to *Agrobacterium*
tumefaciens, confers on the recipient the ability to nodulate the host
legume normally nodulated by the strain from which the sym plasmid was
obtained [Johnston et al.,1978; Hooykaas et al.,1981; Lamb et al.,1982].
In one such sym plasmid of a strain of *R. l.* bv. *viciae*, a region less than
10kb in size was shown to confer the ability to nodulate peas when

transferred to strains of *Rhizobium* lacking a sym plasmid or to
A.tumefaciens [Downie et al.,1983; Govers et al.,1986]. Within this
region, ten genes, *nodL,M,E,F,D,A,B,C,I* and *J* were identified by DNA
sequencing and isolation and charecterization of *nod* mutations [Rossen et
al.,1984; Downie et al.,1985,1987; Evans and Downie, 1986; Shearman et al.,
1986]. The structure and the possible functions of these genes have
recently been reviewed [Rossen et al.,1987]; despite the detailed physical
description of these genes, though, the precise role of only one of them,
nodD, has been established.

It has been shown that in *R.l.* bvs. *viciae* and *trifolii* and also in
R.meliloti [which nodulates alfafa], the only *nod* gene that is actively
transcribed in cells growing in normal growth media is *nodD* [Innes et
al.,1985; Mulligan and Long, 1985; Rossen et al.,1985; Shearman et
al.,1986]. However, when the rhizobia were exposed to the root exudate of
their hosts, transcription of the other *nod* genes was induced, this
activation being dependant on the presence of the regulatory *nodD* gene
[Mulligan and Long, 1985; Rossen et al.,1985; Innes et al.,1985]. It was
established that the inducer molecules in the root exudates are particular
flavones or flavanones; Peters et al.,[1986] showed that the flavone
luteolin was the most potent inducer of transcription of *nod* genes of
R.meliloti which was present in the root exudate of alfafa and Redmond et
al.,[1986] identified the inducer flavone 7,4'-dihydroxy flavone in the
exudate of clover seedlings which activated the *nod* genes of *R.l.* bv.
trifolii. Similarly, Firmin et al.,[1986] showed that several flavones and
flavanones, all of which had hydroxyl or glycosidic substitutions in the 7
position of the A ring and the 4' position of the B ring were inducers of
the *nod* genes of *R.l.* bv. *viciae*. These inducers are remarkably potent; at
concentrations as low as 200nM, significant increases in transcription of
nod genes are observed [Firmin et al.,1986; Redmond et al.,1986].

As with other *nod* genes [see Rossen et al.,1987], there is strong
conservation in the sequences of the *nodD* genes of different *Rhizobium*
biovars and species [Shearman et al.,1986; Egelhoff et al.,1985; Schofield
and Watson,1986; Gottfert et al.,1986]. This conservation in sequences is
reflected in the fact that, in some cases at least, the *nodD* genes of
different rhizobia are funtionally equivalent. For example, the Nod⁻
defect on clover of a *nodD* mutant strain of *R.l.* bv. *trifolii* was corrected

by the cloned *nodD* gene of *R. meliloti* [Fisher et al.,1985].   Also, certain
flavonoid molecules [eg. luteolin] have been shown to be inducers for
different *Rhizobium* species [Peters et al.,1986; Firmin et al.,1986].
However, more recent studies have shown that *nodD* genes of different
rhizobia do not have identical "sensitivities" to different inducer
flavonoids.   For example, the flavanone hesperitin is a very potent inducer
for *R.l.* bv. *viciae* [Firmin et al.,1986] but is relatively inefficient in
R. meliloti [Peters et al.,1986].   Such differences can, in fact determine
the host range of a particular strain.   Horvath et al.,[1987] showed that
the ability of a wide host-range strain [NGR234] to nodulate the tropical
legume *Macroptillium* [siratro] was due to its possession of a version of
nodD which, when transferred to *R. meliloti*, conferred on the recipient
strain the ability to nodulate this tropical legume.   Presumably the
exudate of siratro contains an inducer which is "recognized" by the *nodD*
gene product of the wide host range strain but not by that of *R. meliloti*.
        Another significant difference in the status of *nodD* genes in
different rhizobia is that, whereas in *R.l.* bv. *viciae* there is only a
single copy of *nodD*, in *R. meliloti* there are three copies of the gene, all
of which are on the sym plasmid of this species [Gottfert et al.,1986].
This duplication of *nodD* in *R. meliloti* is probably responsible for the fact
that the initially detected *nodD* mutations in this species had little
effect on nodulation ability [Gottfert et al.,1986] whereas *nodD* mutations
in *R.l.* bv. *viciae* and *R.l.* bv. *trifolii* completely abolish nodulation of
peas and clover respectively [Downie et al.,1985; Schofield and Watson,
1986].   One other major difference in the properties of *nodD* in different
rhizobia is that in *R.l.* bv.. *viciae*, this gene is autoregulatory;
itrepresses its own transcription [Rossen et al.,1985] but no such effect
was found for *nodD* of *R. meliloti* [Mulligan and Long,1985].
        Upstream of transcriptional units activated by *nodD* plus inducer
flavonoids is a conserved sequence, the so-called *nod*-box that has been
implicated in the control of transcription of the genes that it precedes
[Rostas et al.,1986; Schofield and Watson, 1986; Scott,1986; Shearman et
al.,1986]. It has not been formally proved that the product of *nodD* [NodD]
binds to the *nod*-box, but, given the similarity of NodD [Shearman et
al.,1986] to the the product of *araC*, an *E.coli* gene whose product binds to
regulatory sequences of DNA and which is involved both in autoregulation

and, in the presence of arabinose, in the activation of genes for the catabolism of the sugar [Lee et al.,1981], it seems likely that NodD does indeed bind to the *nod*-box. Also, recent evidence from this laboratory supports this conclusion [Hong et al.,1987; Burn et al.,1987; see below] These two studies also indicated that NodD binds directly with the inducer flavonoids.

One final point concerning the regulation of *nod* genes in *Rhizobium*. Plants not only exude molecules that, *via nodD*, activate transcription of *nod* genes, they also produce related molecules that antagonize this induction. Certain flavonols, iso-flavonoids and one coumarin have been shown to inhibit induction of *nod* gene transcription by root exudate or defined flavonoid inducers [Firmin et al.,1986; Djordjevic et al.,1987]. It will be important to establish if the host legume determines the level of *nod* gene transcription by modulating the concentrations of inducer and "anti-inducer" molecules and if so, how.

In addition to the *nod* genes on the sym plasmid, strains of *Rhizobium* require other genetic material to form nodules. One class of non-nodulating mutants that we have isolated are defective in the production of the high molecular weight acidic exopolysaccharide. The genes responsible for the synthesis of this polymer are not on the sym plasmid [strains that lack the sym plasmid have a normal mucoid appearance]. However, the effects on nodulation ability of mutations that abolish EPS synthesis depend on the particular host-range specified by individual sym plasmids. For example, Borthakur et al.,[1986] introduced the same mutant allele that abolished EPS production into two strains of *R.leguminosarum* which differed only in the identity of their sym plasmids. One contained pRP2JI, a plasmid that specifies the ability to nodulate *Phaseolus*, beans and the other contained pRL1JI, a sym plasmid that confers the ability to nodulate peas and *Vicia* [see above]. The *pss* [polysaccharide synthesis] mutations in both biovars caused a non-mucoid phenotype, but, whereas the *pss* mutations had no observable effect on nodulation or nitrogen fixation on *Phaseolus* by biovar *phaseoli*, the non-mucoid strains of bv. *viciae* failed to nodulate peas or *Vicia* [Borthakur et al.,1986]. It is clear, then, that the requirement for EPS synthesis for nodulation depends on the particular host legume.

Thus, in *R.l.* bv. *viciae* we had isolated two distinct types of Nod⁻ mutants, one of which failed to nodulate because of a deficiency in EPS synthesis and the other which lacked or had mutations in its sym plasmid. In this paper, the effects of co-inoculating peas with these two different tpes of Nod⁻ mutants are described. Also, the effects on symbiotic nitrogen fixation of point mutations in *nodD* which alter its regulatory properties are presented.

RESULTS and DISCUSSION

Mutations in *nodD* which affect its regulatory properties

Plasmid pIJ1518 [Rossen et al.,1985] contains the *nodD* gene of the *R.l.* bv. *viciae* sym plasmid pRL1JI, cloned in the wide host-range vector pKT230. DNA of pIJ1518 was treated *in vitro* with hydroxylamine and derivatives were sought which were defective in autoregulation or in flavonoid-dependent activation of *nodABC*. Also, we screened for *nodD* mutations that activated the other *nod* genes even in the absence of inducer molecules. The strategy used for the isolation of these mutant derivatives was to mobilize the mutagenized derivatives of pIJ1518 into *Rhizobium* strains containing *nodD-lacZ* or *nodC-lacZ* fusion plasmids and, on medium containing X-gal, examine colonies of the transconjugants which, with the *nodD*-lacZ fusion recipient, were blue [indicative of the failure of the mutant *nodD* to auto-regulate] or, in the case of the *nodC-lacZ* fusion recipient were white on media supplemented with pea root exudate or with defined inducer flavonoids [due to the loss of activation]. In addition, mutations in *nodD* which allowed the gene to activate the *nodC-lacZ* fusion even in the absence of inducers were isolated [these were identified by plating derivatives of the *Rhizobium* strain containing the *nodC-lacZ* fusion into which the mutagenized *nodD* plasmids had been introduced on medium containing X-gal but lacking inducer molecules and examining the colonies for any that were blue]. These screens allowed the isolation of four Classes of mutant forms of *nodD* [Burn et al.,1987].

ClassI; these mutations caused a deficiency both in auto-regulation and, in the presence of flavonoid inducers, in the induction of the *nodC-lacZ* fusion.

ClassII; these mutations abolished the auto-regulatory properties of *nodD*, but left unimpaired its ability to activate the other *nod* genes.

ClassIII; these mutations conferred the "reciprocal" phenotype of those of ClassII in that they did not affect autoregulation but caused a specific defect in the activation of the *nodC-lacZ* fusion.

ClassIV; two mutant forms of *nodD* which activated the *nodC-lacZ* fusion in the absence of inducer molecules were isolated. These mutations, which were probably clonal [both had the same base pair change], were in the region of *nodD* which specified the carboxy- terminal region of the gene product. Interestingly, strains with these mutant forms of *nodD* were impaired in their ability to nodulate [approximately 10 nodules per plant compared to 90 nodules on peas inoculated with wild-type strains]. Also, the nodules that formed failed to fix nitrogen, indicating that constitutve, high level expression of *nod* genes is detrimental to nitrogen fixation by bacteroids. Although strains with these ClassIV mutants of *nodD* expressed the other *nod* genes in the absence of flavonoid inducers, when such molecules were added to the cells, there was "hyper-induction" of the *nodC-lacZ* fusion showing that the product of these mutant forms of *nodD* still responded to the inducers. It was striking, moreover, that molecules such as kaempferol, genestein and daidzein, which in wild-type strains of *R.1.* bv. *viciae* are "anti-inducers" [Firmin et al.,1985], not only failed to antagonize induction of the transcription of *nodC* by the flavone or flavanone inducers but actually acted as inducers for expression of the *nodC* and *nodF* genes. The fact that a single mutation in *nodD* affected the response to both inducer and anti-inducer molecules provides strong circumstantial evidence that both types of molecule interact directly with NodD and that the anti-inducers act by preventing the binding of the inducers to this protein [Burn et al.,1987].

Further studies on these mutant forms of *nodD* have provided evidence that NodD binds to DNA and that this binding is responsible for at least one of the regulatory properties of this gene [Hong et al.,1987]. A fragment of DNA that spans the inter-genic region between *nodD* and the divergently transcribed *nodA* [Shearman et al.,1986] was end-labelled and was exposed to extracts of various *Rhizobium* strains before being subjected to electrophoresis and autoradiography. When the extract was from a strain containing a copy of *nodD*, a retarded band, representing a DNA-protein

190

complex was observed on the autoradiograph [Hong et al.,1987]. It seemed
likely that this complex contains the *nodD* gene product since it was formed
when the extract was obtained from a strain of *E.coli* containing the cloned
nodD of *R.l.* bv. *viciae* [Hong et al.,1987].

Co-inoculation of peas with two different types of Nod⁻ mutants of *R.l.* bv. *viciae*

As described in the INTRODUCTION, mutants of *R.l.* bv. *viciae* which
fail to make the acidic polysaccharide have been isolated and have been
found to be unable to nodulate peas [Borthakur et al.,1986]. It was of
interest, therefore, to establish if these EPS⁻ mutant forms could be
"rescued" into nodules when used as one of the strains in a mixed
inoculation experiment in which the other strain [strain 8401; Lamb et
al.,1982] could make EPS but failed to nodulate because it lacked a sym
plasmid. Peas inoculated with such a pair of Nod⁻ strains formed normal
numbers of nodules. When the bacteria were isolated from such nodules, it
was found that they comprised exclusively the strain that lacked the sym
plasmid, indicating that a strain lacking the *nod* genes on the sym plasmid
can be efficiently helped into the nodule by the strain which contains a
sym plasmid but which fails to nodulate because of its defect in EPS
production. This observation emphasizes the conclusion that the *nod* genes
on the sym plasmid are involved only in the early stages in the infection
process and that, for the development of nitrogen-fixing nodules, other
Rhizobium genes are essential. Future studies will be aimed at identifying
these genes and also at deterimining the precise molecular functions of
the plasmid-linked genes that are also so clearly involved in the
nodulation process.

ACKNOWLEDGMENTS

This work was supported by the Agricultural and Food Research Council.
G.F. Hong was in receipt of a travel grant from the Rockefeller Foundation
and J.L.B. and J.W.L. were respectively funded by the Agricultural Genetics
Company, Cambridge and the John Innes Foundation.

REFERENCES

Borthakur, D., Barber, C.E., Lamb, J.W., Daniels, M.J., Downie, J.A. and
    Johnston, A.W.B. 1986. A mutation that blocks exopolysaccharide
    synthesis prevents nodulation of peas by *Rhizobium leguminosarum* but
    not of beans by *R.phaseoli* and is corrected by cloned DNA from
    Rhizobium or the phytopathogen *Xanthomonas*. Mol. Gen. Genet., 203,
    320-323.
Burn, J.L., Rossen, L. and Johnston, A.W.B. 1987. Four classes of
    mutations in the *nodD* gene of *Rhizobium leguminosarum* biovar
    viciae that affect its ability to autoregulate and/or to activate
    other *nod* genes in the presence of flavonoid inducers. Genes and
    Dev., [in the press].
Djordjevic, M.A., Redmond, J.W., Batley, M. and Rolfe, B.G. 1987. Clovers
    secrete specific phenolic compounds which either stimulate or
    repress *nod* gene transcription in *Rhizobium trifolii*. EMBO J., 6,
    1173-1179.
Downie, J.A., Hombrecher, G., Ma, Q.S., Knight, C.D., Wells, B. and
    Johnston, A.W.B. 1983. Cloned nodulation genes of *Rhizobium
    leguminosarum* determine host range specificity. Mol. Gen. Genet., 190,
    359-365.
Downie, J.A., Knight, C.D., Johnston, A.W.B. and Rossen, L. 1985.
    Identification of genes and gene products involved in the nodulation
    of peas by *Rhizobium leguminosarum*. Mol. Gen. Genet., 198, 278-282.
Downie, J.A., Surin, B.P., Evans, I.J., Rossen, L., Firmin J.L., Shearman,
    C.A. and Johnston, A.W.B. 1987. Nodulation genes of *Rhizobium
    leguminosarum*. in Molecular Genetics of Plant-Microbe Interactions
    [eds. Verma, D.P.S. and Brisson, N.] Martinus Nijhoff, Publishers, pp
    225-228
Egelhoff, T.T., Fisher, R.F., Jacobs, T.W., Mulligan, J.T. and Long, S.R.
    1985. Nucleotide sequence of *Rhizobium meliloti* 1021 nodulation genes;
    nodD is transcribed divergently from *nodABC*. DNA 4, 241-248
Evans, I.J. and Downie, J.A. 1986. The *nodI* gene product of *Rhizobium
    leguminosarum* is closely related to ATP-binding bacterial transport
    proteins; nucleotide sequence of *nodI* and *nodJ*. Gene, 43, 95-101.
Firmin, J.L., Wilson, K.E., Rossen, L. and Johnston, A.W.B. 1986. Flavonoid
    activation of nodulation genes in *Rhizobium* reversed by other
    compounds present in plants. Nature, 324, 90-92.
Fisher, R.F., Tu, J.K. and Long, S.R. 1985. Conserved nodulation genes in
    R.meliloti and *R.trifolii*. Appl. Environ. Microbiol., 49, 1432-1430
Gottfert, M., Horvath, B., Kondorosi, E., Putnocky, P., Rodriguez-
    Quinones, F and Kondorosi, A. 1986. At least two *nodD* genes are
    necessary for efficient nodulation of alfafa by *Rhizobium meliloti*. J.
    Mol. Biol., 191, 411-420
Govers, F., Moerman, M., Downie, J.A., Hooykaas, P.J.J., Franssen, J., van
    Kammen, A. and Bisseling, T. 1986. *Rhizobium nod* genes are involved in
    in inducing an early nodulin gene. Nature 323, 564-566.
Hong, G.F., Burn, J.L. and Johnston, A.W.B. 1987. Analysis of *nod* gene
    regulation in *Rhizobium*, in Proceedings of N.A.T.O. advanced study
    workshop on Plant Molecular Biology, [von Wettstein, D. and Chua,
    N.H., eds.] Academic Press [in press].
Horvath, B., Bachem, C.W.B., Schell,J. and Kondorosi, A. 1987. Host-
    specific regulation of nodulation genes in *Rhizobium* is mediated by a
    plant-signal, interacting with the *nodD* gene product. EMBO J., 6, 841-

848

Hooykaas, P.J.J., van Brussel, A.A.N., den Dulk-Raas,H., van Slogteren,
     G.M.S. and Schilperoort, R.A. 1981. Sym plasmid of *Rhizobium trifolii*
     expressed in different rhizobial species and Agrobacterium
     tumefaciens. Nature, 291, 351-353.
Innes,R.W., Kuempl,P.L., Plazinski,J., Canter-Cremers,H. and Rolfe,B.G.
     1985. Plant factors induce expression of nodulation and host-range
     genes in *Rhizobium trifolii.* Mol. Gen. Genet., 201, 426-432
Johnston, A.W.B., Beynon, J.L., Buchanan-Wollaston, A.V., Setchell, S.M.,
     Hirsch, P.R. and Beringer, J.E. 1978. High frequency transfer of
     nodulating ability between strains and species of *Rhizobium.* Nature,
     276, 634-636.
Lamb, J.W., Hombrecher, G. and Johnston, A.W.B. 1982. Plasmid-determined
     nodulation and nitrogen fixation abilities in *Rhizobium phaseoli.* Mol.
     Gen. Genet. 186, 449-452.
Lee, N., Wilcox, G., Gielow, W., Cleary, P. and Engelsberg, E. 1974. *in*
     vitro activation of the transcription of *araBAD* operon by *araC*
     activator. Proc. Nat'l. Acad. Sci. U.S.A., 71, 634-638.
Mulligan, J.T. and Long, S.R. 1985. Induction of *Rhizobium meliloti nodC*
     expression by plant exudate requires *nodD.* Proc. Nat'l. Acad. Sci.
     U.S.A., 82, 6609-6613
Peters, N.K., Frost, J.W. and Long, S.R. 1986. A plant flavone, luteolin,
     induces expressin of *Rhizobium* nodulation genes. Science, 233, 977-
     980.
Redmond, J.W., Batley, M., Djordjevic, M.A., Innes, R.W., Kuempel, P.L. and
     Rolfe, B.G. 1986. Flavones induce expression of nodulation genes in
     Rhizobium. Nature, 323, 632-635
Rossen, L., Johnston,A.W.B. and Downie, J.A. 1984. DNA sequence of the
     Rhizobium leguminosarum nodulation genes *nodAB* and *C* required for
     root hair curling. Nucl. Acids. Res. 12, 7123-7134.
Rossen, L., Shearman, C.A., Johnston, A.W.B. and Downie, J.A. 1985. The
     nodD gene of *Rhizobium leguminosarum* is autoregulatory and in the
     presence of plant exudate induces the *nodA,B,C* genes. EMBO J., 4,
     3369-3373.
Rossen, L., Davis, E.O. and Johnston, A.W.B. 1987. Plant-induced expression
     of *Rhizobium* genes involved in host specificity and early stages of
     nodulation TIBS, [in the press]
Rostas, K., Kondorosi, E., Horvath, B., Simoncsits, A. and Kondorosi, A.
     1986. Conservation of extended promoter regions of nodulation genes in
     Rhizobium. Proc. Nat'l. Acad. Sci. U.S.A., 83, 1757-1761.
Schofield, P.R. and Watson, J.M. 1986. DNA sequence of *Rhizobium trifolii*
     nodulation genes reveals a reiterated and potentially regulatory
     sequence preceding *nodABC* and *nodFE.* Nucl. Acids Res., 14, 2891-
     2903.
Scott, K.F. 1986. Conserved nodulation genes from the non-legume symbiont
     Bradyrhizobium Sp. parasponia. Nucl. Acids Res., 14, 2905-2919.
Shearman, C.A., Rossen, L., Johnston, A.W.B. and Downie, J.A. 1986. The
     Rhizobium leguminosarum nodulation gene *nodF* encodes a polypeptide
     similar to acyl carrier protein and is regulated by *nodD* plus a factor
     in pea root exudate. EMBO J. 5, 647-652.

THE ROLE OF NODULINS IN ROOT NODULE DEVELOPMENT

F. Govers and T. Bisseling

Department of Molecular Biology,
Agricultural University, De Dreijn 11,
6703 BC Wageningen, The Netherlands.

ABSTRACT

In leguminous root nodules and during the formation of these nodules several plant genes, the so-called nodulin genes, are specifically expressed. The nodulins encoded by these genes appear differentially during the development of the symbiotic interaction with Rhizobium. Early nodulin genes are expressed during the stage that nodule structures are formed and the nodulins encoded by these genes can have functions in the infection process or nodule morphogenesis. Late nodulins are first detectable when nitrogen fixation starts and they can have a role in supporting the nitrogen fixing bacteroids or in the assimilation of fixed nitrogen. The expression of nodulin genes is specifically induced by Rhizobium. The nodulation genes of Rhizobium appear to be involved in the induction of at least one early nodulin gene. The concentration of late nodulin mRNAs varies in different types of ineffective nodules suggesting that Rhizobium also has a role in regulating the expression level of nodulin genes.

NODULINS AND NODULIN GENES

The symbiotic association of legumes and bacteria of the genera Rhizobium and Bradyrhizobium results in the formation of specialised organs (root nodules) in which the bacteria are capable of fixing atmospheric nitrogen. The nodule formation and nitrogen fixation depends on complex genetic interactions between the two organisms. This paper deals with the nodulin genes, the genes of the host plant which are specifically expressed during symbiosis. Classical genetic analyses of plant mutants have shown that plant genes play important roles in all steps of the symbiotic process, from the preinfection stage up to and including the assimilation of ammonia in a mature nitro-

193

F. O'Gara et al. (eds.), Physiological Limitations and the Genetic Improvement of Symbiotic Nitrogen Fixation, 193–206.
© 1988 by Kluwer Academic Publishers.

gen fixing root nodule (La Rue et al., 1985). In the last decade
molecular, biochemical and immunological techniques have been
used to identify nodulins, to clone nodulin genes, and to ana-
lyse nodulin gene expression (for reviews see Verma et al.,
1986; Govers et al., 1987a).

Nodulin gene expression has been studied in soybean
(Legocki and Verma, 1980), pea (Govers et al., 1985), vetch
(Moerman et al., 1987), alfalfa (Lang-Unnasch and Ausubel,
1985) and yellow lupin (Stroczycki et al., 1985). These studies
have shown that leguminous root nodules contain at least
twentyfive different nodulins. By following the appearance of
nodulin mRNAs during the formation of pea and soybean nodules,
it was found that the nodulin genes are differentially expressed
(Govers et al., 1985; Gloudemans et al., 1987). Some nodulin
genes are expressed during the stage that nodule structures are
formed, but expression of the majority of the nodulin genes in
pea as well as in soybean starts a few days later, just before
or just after the onset of nitrogen fixing activity in the nodules. These observations have resulted in the classification of
nodulins into two groups, early nodulins and late nodulins. For
each of these two groups nodulin cDNA clones have been isolated
(Fuller et al., 1984; Govers et al., 1987b; Franssen et al.,
1987) and they have proved to be important tools to study the
role of the host plant in root nodule formation and nodule functions.

FUNCTIONS OF NODULINS

Since the expression of the two classes of nodulin genes

coincides with particular stages in the development of the symbiosis, functions of early and late nodulins must be related to
specific events that take place during these stages. Histologi-
cal analyses of nodule development in combination with studies
on the time-course of appearance of nodulins have shown that
early nodulin genes are expressed at the stage that the nodule
structure with all its defining characteristics is being formed.
At that stage late nodulin transcripts are not yet detectable
(Govers et al., 1987b; Gloudemans et al., 1987). Therefore early
nodulins can be involved in nodule organogenesis or the infection process. The expression of late nodulin genes is first detectable when nitrogen fixation starts. Hence late nodulins will
most probably function in establishing and maintaining a proper
environment within the nodule that allows nitrogen fixation and
ammonium assimilation to occur.

EARLY NODULINS

From the small group of early nodulins that have been identified so far only one has been studied in some detail and that
is the soybean early nodulin Ngm-75. By comparing two dimension-
al gels containing the in vitro translation products of RNA isolated from uninfected soybean roots and infected root segments,
Gloudemans et al. (1987) found that Ngm-75 appears as early as 7
days after infection with Bradyrhizobium japonicum. Hybrid released translation experiments showed that the soybean cDNA
clone pENOD2, which was selected as an early nodulin cDNA clone,
encodes Ngm-75 (Franssen et al., 1987). On northern blots pENOD2
cross hybridises strongly with nodulin mRNA from pea (Govers et

196

al., 1986a), vetch (Moerman et al., 1987) and alfalfa (unpublished results) showing that a homologous gene is expressed in
different legumes. The widespread occurrence of the ENOD2-like
gene among legumes and its apparent conservation, suggests a
similar function of the encoded protein in the various symbioses.

In nodules induced on pea roots by an Agrobacterium strain
(LBA 2712) that is cured of its Ti plasmid but harbours a R.
leguminosarum sym plasmid instead, infection threads are observed but intracellular bacteria are not. Such nodules have a central tissue surrounded by cortical cell layers and pheripherally
located vascular bundles. In contrast, in pea nodules induced
by wildtype R. leguminosarum the central tissue consists of only
one cell type: relatively small non-dividing cells that do not
resemble the infected or the uninfected cells found in normal
nodules. In these "empty" pea root nodules the ENOD2 gene is
expressed but not any of the late nodulin genes (Govers et al.,
1986a). On soybean roots inoculated with R. fredii (USDA 257)
the formation of nodule-like structures is induced but in these
structures neither infection threads nor intracellular bacteria
are found. Also in alfalfa nodules induced by R. meliloti mutants that have lost the ability to produce exopolysaccharides,
infection threads and cells containing bacteria are missing. In
both cases ENOD2 mRNA has been detected in the nodule tissue
(Franssen et al., 1987; unpublished results). These observations
indicate that Ngm-75 has a function in nodule morphogenesis and
not in the infection process.

DNA sequence analysis of the cDNA clone pENOD2 (Franssen et

<u>al</u>., 1987) revealed that proline is the major amino acid of the
Ngm-75 nodulin. The amino acid sequence is organised in highly
repetitive units, composed primarily of two predominant
pentapeptide repeats Pro-Pro-His-Glu-Lys and Pro-Pro-Glu-Tyr/His
-Gln. It seems likely that the prolines become hydroxylated <u>in</u>
<u>vivo</u> to yield a hydroxyproline-rich protein. Several classes of
hydroxyproline-rich proteins have been described to occur in
plants (Varner, 1987). The best studied class comprises the cell
wall structural hydroxyproline-rich glycoproteins (HRGPs) or
extensins which also have a highly repetitive structure. In this
case the characteristic repeating unit is Ser-Hyp-Hyp-Hyp-Hyp.
Extensins are found in the cell walls of most dicotyledonous
plants and they have been shown to accumulate upon wounding,
pathogen attack and treatment with oligosaccharide elicitors or
ethylene. Thus they are assumed to be involved in the plant defence mechanism. The structural features of the Ngm-75 nodulin
and the similarities found with the extensins do not prove, but
strongly suggest that Ngm-75 is a cell wall protein. This suggestion is furthermore supported by the finding that the cortex of
the soybean root nodule is one of the richest sources of cell
wall hydroxyproline yet found (Cassab, 1986). Ngm-75 as a cell
wall protein fits in the above deduced role of this early nodulin in nodule formation.

LATE NODULINS

    Among the large group of late nodulins leghemoglobin is the
most abundant and also the best characterised nodulin (reviewed
by Appleby, 1984). Twentyfive percent of the total soluble pro-

tein fraction in the nodule consists of leghemoglobin. Leghemo-globin is a myoglobin-like hemoprotein with a high affinity for oxygen. It is localised in the cytoplasm of the infected cells where it controls the concentration of free oxygen in such a way that the bacteroids are efficiently supplied with oxygen. Thus leghemoglobin supports the functioning of the bacteroids.

Other known late nodulins are involved in the assimilation of reduced nitrogen, for example Ngm-35 which is the second most abundant nodulin in soybean nodules. It has been shown that Ngm-35 is the 33 kDa subunit of n-uricase (or uricase II) (Verma et al., 1986), a key enzyme in the ureide biosynthetic pathway used in soybean to assimilate ammonia. n-Uricase is exclusively found in the peroxisomes of uninfected nodule cells. In soybean seve-ral enzymes involved in the ureide synthetic route are localised here which suggests that there is a metabolic specialisation of uninfected cells. Sequence analysis of the soybean n-uricase cDNA clone did not reveal a signal peptide, so the transport into peroxisomes must reside in the protein itself. Two hydro-phobic domains present in the amino acid sequence may facilitate translocation across the peroxisomal membrane.

Another enzyme which is involved in the assimilation of ammonia is glutamine synthetase. Several glutamine synthetase (GS) isozymes are found in different organs and cell compart-ments of plants. During nodule development a significant incre-ase in GS activity is observed. Sequence analysis of various GS cDNA clones from leguminous plants indicated a very high homolo-gy (70-90%) in the coding sequences within and between species. In contrast, the 5'- and 3'-untranslated regions of the GS cDNA clones have diverged highly, and thus these regions of the cDNA clones have been used to prove the existence of a nodule specific GS gene. An alfalfa n-GS cDNA clone was isolated that could be distinguished from other GS sequences because of a unique sequnce present in the 3'-untranslated region (Dunn et al., submitted). A cDNA clone for the nodule specific GS subunit of French bean was shown to hybridise exclusively to nodule RNA under stringent conditions, thus confirming the existence of a n-GS gene (Cullimore et al., 1984). The amino acid homology between the root and nodule GS proteins agrees with the remarkably similar biochemical proprties of GS isozymes. Therefore, the differential expression of GS genes appears to result in functionally similar enzymes. In nodules a new form might be required for assimilating the large quantities of ammonium produced by the bacteroids. By turning on expression of nodulin genes involved in ammonium assimilation the plant has the ability to boost its assimilation capacity within the nodule cytosol several fold and it seems very likely that other late nodulins also have an enzymatic function in the assimilation pathways.

Besides nitrogen assimilation other metabolic pathways in the nodule may require the presence of late nodulins. For example a soybean nodulin, Ngm-100, has been shown to be the subunit of sucrose synthetase (Thummler and Verma, submitted). Moreover for several enzymes, nodule specific forms that differ in physical, kinetic or immunochemical properties from their root counterparts have been found. Examples include phosphoenolpyruvate carboxylase (Deroche et al., 1983), choline kinase (Mellor et al., 1986), xanthine dehydrogenase (Verma et al., 1986), purine

nucleosidase (Larsen and Jochimsen, 1987) and malate dehydrogenase (Appels and Haaker, submitted). These may be nodulins but
it has not yet been proven that the nodule specific forms are
encoded by genes which are only expressed in the nodule. The
nodule specific forms can also be the result of posttranscript-
ional modifications of genes which are expressed in other
tissues as well.

In soybean at least three nodulins, Ngm-23, Ngm-24 and Ngm-
26, are associated with the peribacteroid membrane. This membrane is the physical and metabolical interface between Rhizobium
and the plant cytoplasm. As such, the involvement of peribacteroid membrane associated nodulins in the symbiosis seems self-
evident. Ngm-24 has a remarkable hydrophobic and repetitive
structure and it contains a signal peptide (Verma et al., 1986).
Secondary structure analysis suggests that the Ngm-24 protein
resides as an alpha-helix on the surface of the peribacteroid
membrane, facing the peribacteroid space, which indicates that
Ngm-24 does not function as a membrane transport molecule. On
the contrary the secondary structure of Ngm-26 protein suggests
that it is a transmembrane protein, embedded in the peribacteroid membrane facing both sides (Fortin et al., 1987). Thus Ngm-
26 may function in the exchange of metabolites between the Rhizobium and the plant. Ngm-23 belongs to a group consisting of
five related nodulins, Ngm-20, Ngm-23, Ngm-26b, Ngm-27 and Ngm-
44 (Jacobs et al., 1987). These nodulins have regions in common
and regions unique to each particular nodulin. Except for Ngm-
26b, they all possess a conserved hydrophobic signal sequence
for membrane translocation. Two other conserved domains are cen-

tered around four cysteine residues. The spatial distribution of these cysteines suggests that most members of this family consist of metal binding polypeptides, that either cross a membrane or are associated with it. For Ngm-23 it has been shown that it is located in the peribacteroid membrane, but the location of the other members of this family has not been established yet. Despite the homology between these nodulins, insertions and deletions result in completely different polypeptides. The functions of these nodulins are still unknown.

INDUCTION AND REGULATION OF NODULIN GENE EXPRESSION

In order to study the mechanisms by which nodulin gene expression is induced we searched for the origin of the signals that are involved. Although the Rhizobium-legume interaction leads to a symbiotic relation it can not be excluded that infection with Rhizobium elicits a defence reaction from the plant. Some of the proteins that are identified as nodulins could be involved in this host defence mechanism implicating that phytopathogenic organisms might just as well be able to induce expression of these genes. To test this hypothesis we analysed the gene expression in pea tumors induced by Agrobacterium tumefaciens which is closely related to Rhizobium, but no nodulin mRNAs were detected.

Once nodules are formed the physiological situation within the nodule cells is different from that in the root cells. The free oxygen concentration is very low and it has been suggested that these microaerobic conditions might cause enhanced activities of enzymes involved in an anaerobic metabolism (De Vries et al., 1980). In pea roots grown in a microaerobic environment the expression of at least 8 genes is specifically induced but none of them are nodulin genes (Govers et al., 1986b). From these observations we concluded that nodulin gene expression is not related to a general host defence mechanism and microaerobiosis alone is not sufficient for induction. Apparently the Rhizobium delivers specific inducing signals.

To investigate what kind of signals from Rhizobium are involved in inducing or regulating nodulin genes, we analysed the nodulin gene expression in root nodules formed by several Rhizobium mutants and engineered Rhizobium and Agrobacterium strains. In ineffective nodules formed on pea by strains that induce the formation of root nodules which morphologically have the same organisation and cell types as wild type nodules, all the nodulin genes are expressed (Govers et al., 1985; 1987b). Such nodules are also induced by strains that have only a 12 kb nod fragment instead of the complete R. leguminosarum sym plasmid. Hence it was concluded that besides the 12 kb nod region, the nif, fix and other putative genes encoded by the sym plasmid are not involved in the induction of nodulin genes. The nod region alone might be involved in the induction although the results do not exclude a role for the Rhizobium chromosome. An Agrobacterium strain (LBA 2712) harbouring a complete R. leguminosarum sym plasmid induces nodules on pea in which expression of the early nodulin gene ENOD2 can be demonstrated (Govers et al., 1986a). Since in pea tumors formed by A. tumefaciens the ENOD2 gene is not expressed, it seems unlikely that the Agrobacterium genome itself carries genes involved in the induction of the ENOD2

gene. Therefore the sym plasmid in strain LBA 2712 must be responsible for the induction and from this sym plasmid only the
nod region is essential. The Rhizobium genes involved in inducing ENOD2 expression are thus the nod genes located on the 12
kb fragment.

Induction of expression of the late nodulin genes is not
observed in empty pea root nodules induced by the Agrobacterium
strain LBA 2712. If the expression fails because the genetic
information of the nodulating strain is not sufficient, then
Rhizobium genes located on the chromosome or on non-symbiotic
plasmids are required for the induction of late nodulin genes.
However an alternative explanation could be that the plant re-
acts in a different manner towards the Agrobacterium strain by
recognising it as a pathogen. If the defence mechanisms interfere with the induction of nodulin gene expression then a function for the chromosomal genes in inducing late nodulin genes is
questionable (Nap et al., 1987).

Although sym plasmid genes other than the nod genes are not
required for the induction of nodulin gene expression they do
regulate in some way the expression of nodulin genes. We determined on Northern blots the RNA levels of one early and several
late nodulins in ineffective pea root nodules. It appeared that
the early nodulin gene ENOD2 is transcribed at the same rate in
ineffective and wildtype nodules but the RNA levels of the late
nodulins were decreased in ineffective nodules (Govers et al.,
1987b). This decrease varied in three types of ineffective nodules and in addition the accumulation level of each nodulin mRNA

was influenced in a different way. This suggests that several

regulating mechanisms must exist at the transcriptional level.

Apparently this regulation depends on the Rhizobium genes that

contribute to the final nitrogen fixing capacity of the root

nodule. By gaining more knowledge of these mechanisms one might

find strategies to improve the efficiency of symbiotic nitrogen

fixation through manipulation of either the Rhizobium genes or

the nodulin genes.

ACKNOWLEDGEMENTS

We would like to thank the members of our research group for their contributions and we greatly appreciate the help of G. Heitkonig for typing the manuscript.

REFERENCES

Appleby, C.A. 1984. Leghemoglobin and Rhizobium respiration. Ann. Rev. Plant Physiol. 35, 443-478.

Cassab, G.I. 1986. Arabinogalactan proteins during the development of soybean root nodules. Planta, 53, 344-354.

Cullimore, J.V., Gebhardt, C., Saarelainen, R., Miflin, B.J., Idler, K.B. and Barker, R.F. 1984. Glutamine synthetase of Phaseolus vulgaris L.: Organ-specific expression of a multigene family. J. Mol. Appl. Genetics, 2, 589-599.

Deroche, M-E., Carrayol, E. and Jolivet, E. 1983. Phosphoenol-pyruvate carboxylase in legume nodules. Physiol. Veg. 21, 1075-1081.

De Vries, G.E., In't Veld, P. and Kijne, J.W. 1980. Production of organic acids in Pisum sativum root nodules as a result of oxygen stress. Plant Sci. Letts. 20, 115-123.

Fortin, M.G., Morrison, N.A. and Verma, D.P.S. 1987. Nodulin-26, a peribacteroid membrane nodulun is expressed independently of the development of the peribacteroid compartment. Nucl. Acids Res. 15, 813-824.

Franssen, H.J., Nap, J.P., Gloudemans, T., Stiekema, W., Van Dam, H., Govers, F., Louwerse, J. van Kammen, A. and Bisseling, T. 1987. Characterisation of cDNA for nodulin-75 of soybean: a gene product involved in early stages of root nodule development. Proc. Natl. Acad. Sci. USA, 84, 4495-4499.

Fuller, F. and Verma, D.P.S. 1984. Appearance and accumulation of nodulin mRNAs and their relationship to the effectiveness of root nodules. Plant Mol. Biol. 3, 21-28.

Gloudemans, T., De Vries, S.C., Bussink, H-J., Malik, N.S.A., Franssen, H.J., Louwerse, J. and Bisseling, T. 1987. Plant

Mol. Biol. 8, 395-403.

Govers, F., Gloudemans, T., Moerman, M., van Kammen, A. and
    Bisseling, T. 1985. Expression of plant genes during the
    development of pea root nodules. EMBO J. 4, 861-867.

Govers, F., Moerman, M., Downie, J.A., Hooykaas, P., Franssen,
    H.J., Louwerse, J., van Kammen, A. and Bisseling, T. 1986a.
    Rhizobium nodulation genes are involved in expression of an
    early nodulin gene. Nature, 323, 564-566.

Govers, F., Moerman, M., Hooymans, J., van Kammen, A. and
    Bisseling, T. 1986b. Microaerobiosis is not involved in the
    induction of pea nodulin gene expression. Planta, 169, 513-
    517.

Govers, F., Nap, J.P., van Kammen, A. and Bisseling, T. 1987a.
    Nodulins in the developing root nodule. Plant Physiol. Bio-
    chem. 25, 309-322.

Govers, F., Nap, J.P., Moerman, M., Franssen, H.J., van Kammen,
    A. and Bisseling, T. 1987b. cDNA cloning and developmental
    expression of pea nodulin genes. Plant Mol. Biol. 8, 425-
    435.

Jacobs, F.A., Zhang, M., Fortin, M.G. and Verma, D.P.S. 1987.
    Several nodulins of soybean share structural domains but
    differ in their subcellular locations. Nucl. Acids Res. 15,
    1271-1280.

Lang-Unnasch, N. and Ausubel, F.M. 1985. Nodule-specific poly-
    peptides from effective alfalfa root nodules and from in-
    effective nodules lacking nitrogenase. Plant Physiol. 77,
    833-839.

Larsen, K. and Jochimsen, B. 1987. Expression of two enzymes in-
    volved in ureide formation in soybean is regulated by
    oxygen. In "Molecular Genetics of Plant-Microbe Interact-
    ions" (Eds., D.P.S. Verma and N. Brisson), Nijhoff,
    Dordrecht, pp.133-137.

LaRue, T.A., Kneen, B.E. and Gartside, E. 1985. Plant mutants
    defective in symbiotic nitrogen fixation. In "Analyses of
    the plant genes involved in the legume-Rhizobium symbiosis.
    OECD, Paris, pp. 39-48.

Legocki, R.P. and Verma, D.P.S. 1980. Identification of "nodule-
    specific" host proteins (nodulins) involved in the develop-
    ment of Rhizobium-legume symbiosis. Cell, 20, 153-163.

Mellor, R.B., Christensen, T.M.I.F. and Wernwr, D. 1986. Choline
    kinase II is present only in nodules that synthesise stable
    peribacteroid membranes. Proc. Natl. Acad. Sci. USA, 83,
    659-663.

Moerman, M., Nap, J.P., Govers, F. Schilperoort, R., van Kammen,
    A. and Bisseling, T. 1987. Rhizobium nod genes are involved
    in the induction of two early nodulin genes in Vicia faba
    root nodules. Plant Mol. Biol. 9, 171-179.

Nap, J.P., van Kammen, A. and Bisseling, T. 1987. Towards nodul-
    in function and nodulin gene regulation. In "Proceedings of
    the NATO Advanced Study Institute for Plant Molecular Biol-
    ogy ". (Ed. D. von Wettstein), Plenum Press, New York, In
    press.

Strozycki, P. Konieczny, A. and Legocki, A.B. 1985. Identificat-
    ion and synthesis in vitro of plant-specific proteins in
    yellow lupin root nodules. Acta Biochim. Polon. 32, 27-34.

Varner, J.E. 1987. The hydroxyproline-rich glycoproteins of
     plants. In "Molecular Biology of Plant Growth Control"
     (Eds., J.E. Fox and M. Jacobs), Alan R. Liss Inc., New
     York, pp. 441-447.
Verma, D.P.S., Fortin, M.G., Stanley, J., Manro, V.P., Purohit,
     S. and Morrison, N. 1986. Nodulins and nodulin genes of
     Glycine max, a perspective. Plant Mol. Biol. 7, 51-61.

SUBJECT INDEX

F. O'Gara et al. (eds.), Physiological Limitations and the Genetic Improvement of Symbiotic Nitrogen Fixation, 207–210.
© 1988 by Kluwer Academic Publishers.

AUTHOR INDEX

F. O'Gara et al. (eds.), Physiological Limitations and the Genetic Improvement of Symbiotic Nitrogen Fixation, 211–212.
© *1988 by Kluwer Academic Publishers.*

LIST OF PARTICIPANTS

Amarger, N......... Laboratoire de Microbiologie des Sols, INRA, 17 Rue Sully, BV 1540, F-21034 Dijon Cedex, France.

Birkenhead, K...... Department of Microbiology, University College, Cork, Ireland.

Brown, M.......... Kerry Co-op. Research Laboratories, Tralee, Kerry, Ireland.

Cannon, F.C....... Biotechnica International Inc., 85 Bolton Street, Cambridge, Massachusetts 02140, USA.

Casella, S........ Universita di Pisa, Istituto di Microbiologia Agraria e Technica, via del Borghetto 80, 56100 Pisa, Italy.

Condon, C......... Department of Microbiology, University College, Cork, Ireland.

Cooper, J.E....... Department of Agricultural and Food Bacteriology, The Queens's University of Belfast, Belfast BT9 5PX, Northern Ireland UK.

Donnelly, D....... Department of Biology, NIHE, Glasnevin, Dublin 9, Ireland.

Drevon, J.J....... Laboratoire de Recherche sur les Symbiotes des Racines, INRA, 9 Place Viala, 36040 Montpellier Cedex, France.

Finnucane, B...... NBST, Shelbourne House, Shelbourne Road, Dublin 4, Ireland.

Fitzmaurice, A-M... Department of Microbiology, University College, Cork, Ireland.

Gosse, G.......... Station de Bioclimatologie, INRA, 78850 Thievwal-Grignon, France.

Govers, F......... Department of Molecular Biology, Agricultural University, De dreijen 11, 6703 BC Wageningen, The Netherlands.

Haaker, H......... Department of Biochemistry, Agricultural University, De dreijen 11, 6703 BC Wageningen, The Netherlands.

Heckmann, M.O..... Laboratoire de Recherche sur les Symbiotes des Racines, INRA, 9 Place Viala, 36040 Montpellier Cedex, France.

F. O'Gara et al. (eds.), Physiological Limitations and the Genetic Improvement of Symbiotic Nitrogen Fixation, 213–215.
© *1988 by Kluwer Academic Publishers.*

Hynes, M.......... Department of Biology, NIHE, Glasnevin, Dublin 9, Ireland.

Johnston, A.W.B.... Institute of Plant Science Research, John Innes Institute, Colney Lane, Norwich NR4 7UH, UK.

Kahn, D........... Laboratoire de Biologie Moleculaire des Relations Plantes-Microorganismes, CNRS-INRA, BP 27, F-31326 Castanet-Tolosan Cedex, France.

Kahn, M.L......... Department of Microbiology, Washington State University, Pullman, Washington 99164, USA.

Lie, T.A.......... Department of Microbiology, Agricultural University, 4 H. van Suchtelenweg, 6703 CT Wageningen, The Netherlands.

Manian, S.S....... Department of Microbiology, University College, Cork, Ireland.

Marques, J.F...... Department Microbiologica, Estacia Agronomica Nacional, Quinta de Marquies, 2780 Oeiras, Portugal.

Mengel, K......... Institute of Plant Nutrition, Justus-Liebig University, Suedanlage, D-6300 Giessen, F.R. Germany.

Minchin, F.R...... Department of Plant and Cell Biology, Welsh Plant Breeding Station, Plas Gogerddan, Aberystwyth SY23 3EB, UK.

Murphy. P.M....... The Agricultural Institute, Johnstown Castle Research Centre, Wexford, Ireland.

Noonan, B.M....... Department of Microbiology, University College, Cork, Ireland.

Nuti, M.P......... Universita degli Studi di Padova, Istituto di Chimica Agraria e Industrie Agraria, via Gradenigo 6, 35131 Padova, Italy.

O'Gara, F......... Department of Microbiology, University College, Cork, Ireland.

Perkins, B.D...... OECD, 2 rue Andre-Pascal, 75775 Paris Cedex 16, France.

Rao, J.R.......... Department of Microbiology, University College, Cork, Ireland.

Reigh, G.......... Department of Biology, NIHE, Glasnevin,

Dublin 4, Ireland.

Ronson, C.W........ Biotechnica International Inc., 85 Bolton Street, Cambridge, Massachusetts 02140, USA.

Rosendahl, L....... Department of Agricultural Research, Riso National Laboratory, DK 4000 Roskilde, Denmark.

Roskothen, P...... Institut fur Pflazenbau und Pflanzenzuchtung, v-Siebold-Str 8, D-3400 Gottingen, F.R. Germany.

Ruiz-Argueso, T.... Department de Microbiologia, ETS de Ingeneros Agronomos, Universidad Politecnica de Madrid, 28040 Madrid, Spain.

Ryle, G.J.A........ AFRC Institute for Grassland and Animal Production, Hurley, Maidenhead, Berkshire SL6 5LR, UK.

Sheehy, J.E....... AFRC Institute for Grassland and Animal Production, Hurley, Maidenhead, Berkshire SL6 5LR, UK.

Streeter, J.G...... Department of Agronomy, Ohio Agricultural Research and Development Centre, The Ohio State University, Wooster, OH 44691, USA.

Wang, Y-P......... Department of Microbiology, University College, Cork, Ireland.

Werner, D......... Botanisches Institut, FB Biologie, Philipps-Universitat, Karl-von-Frisch-str, D-3550 Marburg/L, F.R. Germany.